The ACP-EU Technical centre for Agricultural and Rural Cooperation (CTA) operates under the Lomé Convention between Member States of the European Union and the African, Caribbean and Pacific States.

CTA collects, disseminates and facilitates the exchange of information on research, training and innovations in the spheres of agriculture and rural development and extension for the benefit of the ACP states.

To achieve this, CTA commissions and publishes studies, organises and supports conferences, workshops and seminars, publishes and co-publishes a wide range of books, proceedings, bibliographies and directories, strengthens documentation services in ACP countries, and offers an extensive information service.

Acknowledgements

I would like to acknowledge the contribution that the following persons have made to the production of the book, John Turton, Cliff Wright, Ruth Gatenby, Anne Pearson and Richard Brown for reading through and making comments on various chapters of the book; Chris McKinnell for numerous line drawings; Boni-Consilli K. Korugyendo for some of the information included in the section on scavenger poultry and poultry diseases; Richard Matthewman for data used for some of the diagrams and tables; T. V. Kumar for the results on variable temperature in the section on climate; Dr P. Mbugua of the University of Nairobi for reading the draft and making comments appropriate to his region, and Pierre Leclercq, formerly of Institut d'Elevage et de Médecine Vétérinaire des Pays Tropicaux of Maisons-Alfort, France, for providing advice on practices in Francophone tropical countries; Dr Noël Chabeuf of Institut d'Elevage et de Médecine Vétérinaire des Pays Tropicaux for arranging Dr Leclercq's input and advising on aspects of the book for the same regions, and to Ms Carol Law for typing the manuscript through its many drafts. AJS

Photographic acknowledgements

The author and publishers wish to acknowledge, with thanks, the following photographic sources.

Ardea pp 18 top; 19 top and bottom; 22 bottom; 159 right CTVM (photographer Robert Munro) p 93 Farm Crafts Today (photographer P. Lyth) p 197 bottom Frank Lane pp 14 top; 15 right; 125 bottom NHPA pp 4; 21; 180 bottom Rex Parry pp 16 bottom; 61 Poultry World pp 6 bottom; 15 left; 16 top; 18 bottom; 20; 126; 159 left

The cover photograph is courtesy of NHPA

All other photographs are courtesy of the author

Titles in the Tropical Agriculturalist series

Sheep	ISBN:0–333–52310–5
Pigs	0–333–52308–3
Goats	0–333–52309–1
Dairying	0–333–52313–X
Poultry	0–333–52306–7
Rabbits	0–333–52311–3
Draught Animals	0–333–52307–5
Ruminant Nutrition	0–333–57073–1
Animal Breeding	0–333–57298–X
Animal Health Vol. 1	0–333–61202–7
Animal Health Vol. 2	0–333–57360–9
Warm-water Crustaceans	0–333–57462–1
Livestock Production Systems	0–333–60012–6

Upland Rice	0–333–44889–8
Tea	0–333–54450–1
Cotton	0–333–47280–2
Weed Control	0–333–54449–8
Spice Plants	0–333–57460–5
Cocoa	0–333–57076–6
The Storage of Food Grains and Seeds	0–333–44827–8
Sugar Cane	0–333–57075–8
Maize	0–333–44404–3
Plantain Bananas	0–333–44813–8
Coffee Growing	0–333–54451–X
Food Legumes	0–333–53850–1
Cassava	0–333–47395–7
Sorghum	0–333–54452–8

Other titles published by Macmillan with CTA (co-published in French by Maisonneuve et Larose)

Animal Production in the Tropics and Subtropics	0–333–53818–8
Coffee: The Plant and the Product	0–333–57296–3
The Tropical Vegetable Garden	0–333–57077–4
Controlling Crop Pests and Diseases	0–333–57216–5
Dryland Farming in Africa	0–333–47654–9
The Yam	0–333–57456–7

Land and Life series (co-published with Terres et Vie)

African Gardens and Orchards	0–333–49076–2
Vanishing Land and Water	0–333–44597–X
Ways of Water	0–333–57078–2
Agriculture in African Rural Communities	0–333–44595–3

THE TROPICAL AGRICULTURALIST

Series Editor
René Coste
Formerly President of the IRCC

General editor, livestock volumes

Poultry

Anthony J. Smith

Centre for Tropical Veterinary Medicine
University of Edinburgh

with the collaboration of
Pierre Leclercq
Formerly of the Institut d'Elevage et de Médecine
Vétérinaire des Pays Tropicaux
Maisons–Alfort
France

CTA
MACMILLAN

First published 1990
Reprinted 1991, 1992 (twice), 1993, 1995 (twice)

The Tropical Agriculturalist Series originated under the title *Le Tehnicien d'Agriculture Tropical* published by G. P. Maisonneuve et Larose, 15 rue Victor-Cousin, 75005 Paris, France, in association with the Agency For Cultural and Technical Co-operation based in Paris, France. Volumes in the series in the French language are available from Maisonneuve et Larose.

Published by MACMILLAN EDUCATION LTD
London and Basingstoke
Associated companies and representatives in Accra, Banjul, Cairo, Dar es Salaam, Delhi, Freetown, Gaborone, Harare, Hong Kong, Johannesburg, Kampala, Lagos, Lahore, Lusaka, Mexico City, Nairobi, São Paulo, Tokyo

Published in co-operation with the Technical Centre for Agricultural and Rural Co-operation, P.O.B. 380, 6700 AJ Wageningen, The Netherlands.

ISBN 0-333-52306-7

Printed in Hong Kong

A catalogue record for this book is available from the British Library.

Front cover photograph of a cross-bred fowl in Bolivia

Contents

Preface vi

1 Poultry in context 1

2 Breeds and strains of poultry and their improvement 2

3 The effect of the tropical environment on poultry and possibilities for modification 47

4 Poultry nutrition 65

5 Intensive systems of egg production 115

6 Intensive systems of poultry meat production 138

7 Eggs and egg incubation 148

8 Disease and disease prevention 162

9 Integration of poultry production into agricultural systems in the tropics 179

10 Marketing poultry products 192

11 Maximising profits from poultry 200

Glossary 206

Bibliography 212

Index 215

Preface

In most western countries poultry production is controlled by large-scale producers who use intensive forms of production. This state of affairs has been reached after a period of forty years during which time poultry production has changed from being a backyard enterprise to one composed mainly of well organised, integrated businesses today. Consequently books concerned with poultry production have been written to meet the needs of medium to large-scale producers using intensive techniques. Few, if any, books have attempted to cover the field covered by this book which addresses itself to the problems of poultry production under tropical and sub-tropical conditions. In these conditions the constraints to increased productivity are very different to those in western countries. Apart from the obvious climatic differences there exist chronic problems in obtaining adequate food supplies and at the same time lack of disease control may make intensive production unprofitable or even untenable. This book attempts to address these problems and to see which of the concepts involved in modern poultry production can be adapted to the tropics and which should be avoided. It approaches the problems both from the point of view of the large-scale producers and also the owners of the village scavenging flocks. It concentrates mainly on laying hens and broilers but does not neglect other species, especially ducks, which are very important in some parts of the world. It aims to be readable, informative and to transfer ideas not only from the west to the tropics but also between tropical countries, this latter form of technological transfer being one of the most useful but also one of the most ignored.

ANTHONY J. SMITH

1 Poultry in context

Poultry in one form or another are kept in most areas of the world. There are fewer religious or social taboos associated with them than there are with pigs and cattle, thus products produced by or from poultry provide an acceptable form of animal protein to most people throughout the world, with the exception of strict vegetarians and vegans. Some Hindu vegetarians are willing to consume eggs produced by intensively kept poultry because these eggs have not been fertilized and are therefore not perceived as living creatures. On the other hand, they are unwilling to consume the eggs produced by scavenging poultry flocks because these will almost certainly be fertile. In certain parts of West Africa the consumption of eggs is discouraged or even prohibited.

The actual number of poultry in the world is very large (Table 1.1) and they far exceed in quantity (although not of course in total weight) the total of sheep, goats, pigs, cattle and buffalo.

The contribution that poultry make to the supply of animal protein varies from area to area (Table 1.2).

The consumption of animal protein per head of population is much greater in developed countries than in developing countries. The average availability of animal protein is between 20 and 30 g per day, which is more than the actual requirements of 6–7 g per day. However, according to an FAO survey in 1970 of ninety countries, sixty-two had an average animal protein availability of less than 30 g per day and seventeen countries have 10 g or less. Rwanda is stated to have the smallest amount of animal protein available (3.6 g/head/day). However, even in those countries that apparently have more than enough animal protein to meet everyone's needs these figures almost certainly give an unrealistic picture of the actual situation. It is certain that even in the poorest countries the consumption of animal protein by the richer people will be at the same level as in western countries whilst the poorest members of these communities may receive no animal protein at all. This is true in some countries in South America, where although the average consumption of protein is high, some sections of the population hardly ever consume animal protein.

It can be argued that this situation could be radically changed and the supply of poultry products in poorer countries could be rapidly

Table 1.1 World population of farm animals, birds and humans in selected countries & regions (1987) (head × 10^6)

	Cattle	Buffalo	Camels	Sheep	Goats	Pigs	Chickens *	Man
Europe	128.4	0.36	–	13.8	13.3	187.4	1248	495.8
North & Central America	166.8	0.009	–	18.4	14.3	92.5	1642	413.2
South America	258.4	0.92	–	109.2	20.2	53.2	811	279.1
Asia	388.5	134.2	4.2	329.0	283.4	410.0	3763	2911.4
China	71.3	20.5	0.5	99.1	67.5	344.2	1796	1084.8
India	199.3	74.3	1.3	55.5	105.0	8.8	175	785.5
Africa	181.5	2.5	14.0	195.2	162.5	12.2	781	589.4
Bangladesh	23.5	1.9	–	1.1	10.8	–	69	106.5
Centrally planned	233.9	24.4	1.3	295.2	81.2	506.0	3409	1576.5
Developed	412.4	0.7	0.27	546.9	28.9	347.1	2642	1227.9
Developing	865.3	137.7	18.2	610.8	472.9	492.7	3394	3769.0
World	1277.7	138.4	18.5	1157.6	501.7	839.8	9445	4996.9

FAO (1982)

* More than 90% of the world's poultry flocks consist of chickens

Table 1.2 Relative importance of four sources of animal protein

Region	Animal protein per head daily (g)	Meat (%)	Milk (%)	Eggs (%)	Fish (%)
Far East	9.1	36	25	8	31
Africa	10.8	51	22	4	23
Near East	13.4	43	47	4	6
Latin America	23.4	56	32	4	8
Europe	41.6	44	40	7	9
North America	69.0	53	34	8	5
World	24.0	38	45	6	11

After Kroeske (1972) (taken from Smith 1990)

expanded to meet the needs for animal protein. That this is technically possible is certainly true because poultry are able to adapt to most areas of the world, have a low economic value, rapid generation time and a high rate of productivity. A poultry enterprise can be producing meat within eight weeks and have the first egg produced within eighteen weeks of the first chick being hatched. All the necessary technology and stock can rapidly be deployed from western countries. As a consequence many governments in developing countries have either encouraged or organised the development of large-scale modern poultry enterprises within their countries. This is partly to increase protein supplies but also to diversify the agricultural economy. However this panacea for the instant provision of cheap animal protein has often been more of a burden than an asset as we will see later. Why do governments see poultry as a means of feeding the population and why are poultry regarded as such efficient producers? To understand why, we need to look at the changes that have taken place over the last forty years within the industry especially in developed countries.

Developments within the poultry industry during the last forty years

In the western world before 1930 most poultry were kept in small flocks as a sideline farm enterprise, usually by the farmer's wife, to provide her with some additional money (Fig 1.1). The method of keeping poultry was not too dissimilar to the scavenger poultry flocks seen in developing countries today. In the villages of West Africa, for example, the wife may

Fig 1.1 *Traditional poultry keeping in England*

keep a dozen or so hens for this purpose. Therefore the question arises, can the developments in poultry production that have taken place over the last forty to fifty years in the west be profitably adopted by developing countries? Are they relevant to all developing countries or are they only relevant to developed industrial countries? The changes have involved all aspects of poultry production but most examples given will be limited to laying hens and broilers where the advances are most important because of their wider application.

Nutrition

Our knowledge of poultry nutrition has changed out of all recognition during this period and as a result we probably know more about poultry nutrition than we do about the nutrition of humans. Consequently dietary requirements can be defined in precise terms respect of energy, amino acids, minerals and vitamins. These requirements can be related in terms of type of bird and even climatic variables and a computer can be used to ensure that the diet formulated is produced at the least cost from the foods that are available.

Genetic improvement

Selection and cross-breeding techniques have enabled the production of a laying fowl which will provide up to 280 eggs a year, as against 100

Fig 1.2 *Broilers – well fed at eight weeks of age, weight 2 kg (note development of comb)*

Fig 1.3 *Males of a laying strain at eight weeks, weight 1 kg*

from unimproved poultry. It has led to the production of a meat type chicken that will weigh over 2 kg at seven weeks of age having been fed on only 4 kg of a balanced diet. This compares very favourably to the figures achieved in the 1930's of 2 kg in fourteen weeks on 10 kg of food. From the domesticated *Gallus domesticus* we have produced two types of bird, one for egg production and one for meat production (Figs 1.2, 1.3 and 1.4). This divergency has not been entirely beneficial because whereas in the past cockerels were fattened and hens were kept for egg production, now the cockerels of egg producing strains are killed on the day they hatch whilst great difficulties are encountered when hens of broiler strains (broiler breeders) are kept for egg production. Only 140 eggs per hen per year are produced by these birds compared with 280 from egg laying strains. In addition broiler breeders have to be fed with

Fig 1.4 *Young chickens showing interaction of genetics and nutrition. Left to right: layer type, low protein diet; broiler type, low protein diet; layer type, high protein diet; broiler type, high protein diet.*

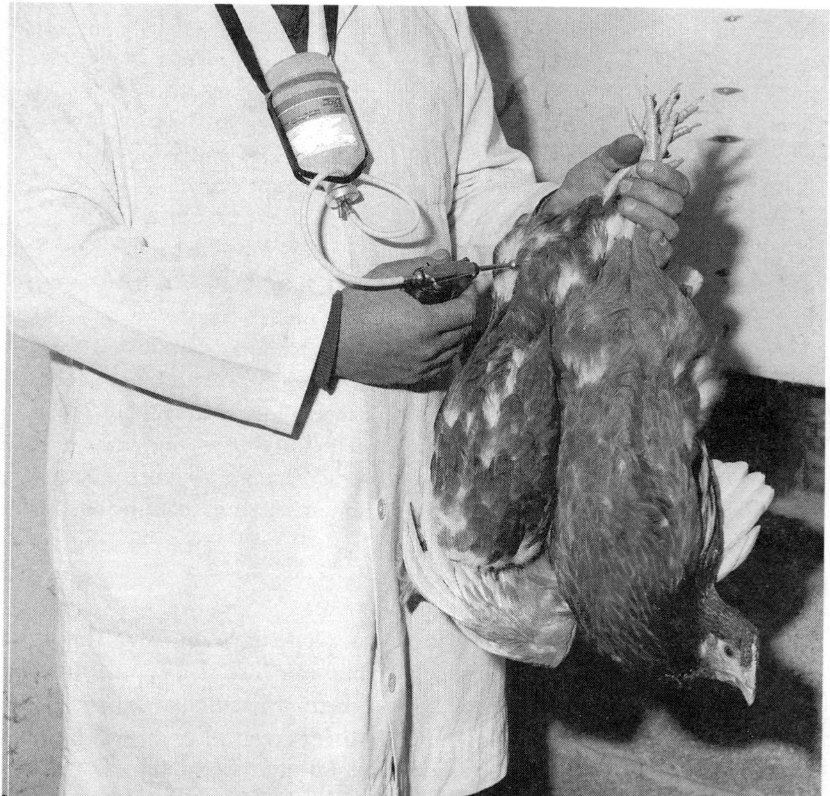

Fig 1.5 *Vaccination by injection*

great precision if they are not to become too fat. Both the broiler and laying strains have been made more productive by utilizing the hybridising techniques developed by maize breeders in the early part of this century.

Disease control

Before the last war the control of disease was more often than not on an individual bird basis, but today the control of disease in whole flocks is the norm. Systems of disease control in flocks now range from isolation (only birds of one age are kept on one particular farm), through flock vaccination for Marek's and Newcastle disease (Fig 1.5) to flock treatment with prophylactics against coccidiosis. The development of disease control in populations has enabled large numbers of birds to be kept in close proximity and has, as a consequence, made intensive poultry management in cages or on deep litter an acceptable method of management.

Management

Not until the first successful incubator in the western world was developed was large-scale poultry production a practical possibility. Although incubators have been known since the time of the Pharaohs and traditional incubators are still found in parts of the Nile Delta and in Indonesia, the first successful one in the west was not developed until 1870.

The other major management change that revolutionised poultry production in the west was the development of vent sexing of chicks. This development, which occurred before the Second World War, enabled male chicks to be identified within twenty-four hours of hatching from the egg. In specialised egg producing strains, the males can be instantly discarded, removing the unnecessary cost of rearing them to ten weeks or so when their sex would normally become evident. Another change was the use of intensive housing and automatic feeding, watering and cleaning out which has reduced the labour requirements so that one man can now manage 30 000 layers or 100 000 broilers. Finally, the use of artificial lighting has enabled hens to maintain egg production all the year round.

Organisation

The final development that has taken place has been in the actual organisation of the poultry industry. Whereas historically all the stages of production were controlled by different individuals or companies, there has been a tendency over the past thirty years for one firm (normally either a poultry breeding company or a food manufacturing company) to take over the whole industry from start to finish. This type of development is known as **vertical integration** and it is seen in its most developed form in the broiler industry as can be seen from the following example. The broiler industry is made up of a number of distinct units as follows.

- Unit 1 – The production of elite meat type breeding stock
- Unit 2 – Broiler hatching egg production
- Unit 3 – Broiler hatcheries
- Unit 4 – Broiler growing
- Unit 5 – Feed compounding
- Unit 6 – Broiler packing
- Unit 7 – Broiler sales (wholesaler and retailer)
- Unit 8 – Broiler sales promotion

Of these units, Units 2 and 6 would normally be under the control of one company, possibly even Units 1 to 8. The advantage of this degree of integration is that every part of the industry can be closely monitored and supplies regulated to meet demands both in terms of quantity and quality.

The place of poultry in food producing systems in developing countries

The developments that have taken place over the last forty years in the poultry industry have been possible only because for most of that period there has been a surplus of food grains in western countries. Poultry are food converters rather than food producers and in the conversion process considerable amounts (over 70%) of both energy and protein that could have been used for human consumption are lost; the loss being far greater when low producing rather than high producing animals are used. Therefore, although the animal protein produced by poultry in the form of eggs and meat has a high nutritional value its production results in a considerable burden being placed on the overall food supplies of the country in which the poultry products are being grown (Tables 1.3 and 1.4).

Thus if adequate supplies of grain do not exist within a country, intensive forms of poultry production cannot be practised. However, as people become richer so their demand for animal protein increases (Table 1.5).

These changes do not reflect a change in nutritional requirements with growing wealth but a desire for foodstuffs which are perceived to be 'better'! The demand for more animal protein reflects a want rather than a need and should be viewed alongside other 'wants' associated with increased wealth such as TV sets and motor cars. According to FAO figures an average adult human needs about 65 g of protein a day, of which only 10% needs to be of animal origin. It can be seen from Table 1.5 that the over-supply in western countries is extreme and may in fact

Table 1.3 The efficiency of food energy conversion of high and low producing domestic farm animals

	Daily M.E.* intake (MJ)	Daily yield of M.E.* by humans (MJ)	Food M.E.* recovered for human consumption (%)
Laying hen (250 eggs/annum)	1.10	0.27	24.5
Laying hen (50 eggs/annum)	0.710	0.075	10.6
Dairy cow (Holstein) (8 000 litres/annum)	172.0	63.0	36.6
Dairy cow (2 000 litres/annum)	81.0	15.7	19.4
Beef steer (gaining 1.5 kg/day)	110.0	18.6	16.9
Beef steer (gaining 0.23 kg/day)	51.0	3.1	6.1

* Metabolizable energy

Table 1.4 Efficiency of protein utilisation by various classes of farm livestock at two levels of output

	Annual crude protein consumed (kg)	Annual crude protein yield (kg)	Food protein recovered for human consumption (%)
Laying hen (250 eggs/annum)	6.6	1.71	26.0
Laying hen (50 eggs/annum)	5.1	0.34	6.7
Dairy cow (8 000 litres/annum)	650.0	260.0	40.0
Dairy cow (2 000 litres/annum)	306.0	65.0	21.0
Beef steer (gaining 1.5 kg/day)	252.0	33.0	13.0
Beef steer (gaining 0.23 kg/day)	124.0	5.5	4.4

A.J. Smith, *Intensive Animal Production in Developing Countries* (BSAP, Thames Ditton, 1981) p. 468.

NB These figures probably give an overestimate of the proportion of protein returned for human consumption as no allowance has been made for the protein cost of keeping breeding herds and flocks.

Table 1.5 Changes in daily per capita consumption of animal and vegetable protein (g) by populations in different regions of the world

	Vegetable protein		Animal protein		Total protein	
	1961–63	1984–86	1961–63	1984–86	1961–63	1984–86
Asia	41.5	48.5	6.9	12.2	48.4	60.7
Africa	44.5	44.8	10.9	12.6	55.4	57.4
Latin America	39.0	37.6	25.0	28.6	64.0	66.2
Europe	44.7	42.7	41.8	58.8	88.4	101.5
USA	33.2	35.6	67.3	70.9	100.5	106.5
World	44.0	46.2	21.3	24.1	65.3	70.3
Estimated daily requirement	58		7		65	

FAO Yearbook (1987)

lead to health problems. Although certain members of the population, such as lactating women and young children, may need a greater proportion of animal protein than the rest of the population, in fact in most countries where animal protein is short more goes to those who need it least, the adult males.

During the last twenty years, and particularly in the last ten, many countries have adopted intensive poultry production as a means of bridging the protein 'gap'. Intensively kept poultry have been seen as a way of rapidly increasing animal protein supplies for rapidly growing urban populations. However what often seems to be forgotten is that the building of a broiler house or a laying house together with a food factory and the necessary processing plants are the easy parts of the operation. The major problem arises when scarce foreign exchange has to be used to import poultry food, and to a lesser extent, vaccines and drugs. To produce 1 kg of broiler carcass requires 2–3 kg of balanced diet and to feed a laying bird throughout the year requires up to 50 kg of a similar diet. The foodstuffs used for poultry food are often of a quality that could be fed directly to humans, so if food is in short supply the effect of intensive poultry keeping is the turning of food that poor people can afford into a much smaller quantity of food that only rich people can afford but almost certainly do not need. Political consideration apart it makes sense for countries to adopt intensive poultry production when one or more of the following conditions prevail.

- The country produces a large surplus of vegetable foodstuffs over and above the needs of the human population.
- The country has a large manufacturing base which brings it sufficient foreign exchange to purchase animal foodstuffs.
- The country has an exportable commodity (such as oil) which it can exchange for animal foodstuffs.

If none of these conditions prevail the intensive poultry industry will become a liability rather than an asset and the government would be better advised to pursue a policy of encouraging scavenger poultry production.

2 Breeds and strains of poultry and their improvement

Species and breeds

Many species, breeds and strains of poultry are used in the service of man. Some species are of limited importance such as the Muscovy duck, guinea fowl, goose and the ostrich. Others, and particularly *Gallus domesticus*, have assumed a worldwide importance. The relationships between the various species are given in Fig 2.1.

Domestic chickens

The genus which includes domestic chickens (*Gallus domesticus*) also includes four wild species; *Gallus gallus*, the red jungle fowl of south-east Asia, *Gallus lafayettei*, the Sri Lanka jungle fowl, *Gallus sonnerati*, the grey jungle fowl found in south-west India and *Gallus varius*, the Javan jungle fowl with green plumage. The domestic fowl most closely resembles the red jungle fowl and readily breeds with it producing fertile hybrids, and at one time the red jungle fowl was thought to be the sole ancestor of all breeds of domestic chicken. Conventional wisdom is that all the other three wild species (*G. lafayettei*, *G. sonnerati* and *G. varius*) interbreed with *G. gallus* and that domestic stock owe some of their inheritance to all of these species. From these various species have arisen domestic hens of various types (Table 2.1). Some, known as the 'fancy breeds', are of little direct commercial value. These fancy breeds do not come within the scope of this book but some such as the Cornish Rock and the White Rock have been very important contributors to the strains that now produce our modern strains of broiler. It is very important that these breeds are maintained in the future as a 'gene bank' because they may contain other useful genes that could be exploited commercially.

At the present time two very different types of *Gallus domesticus* are exploited, one for egg production (Figs 2.2 and 2.3) and one for meat production, whereas previously dual-purpose breeds such as the New Hampshire and Light Sussex were used for both purposes. In today's economic climate in developed countries dual-purpose breeds are regarded as inefficient, producing neither meat nor eggs very efficiently. However in developing countries dual-purpose birds can be very useful especially

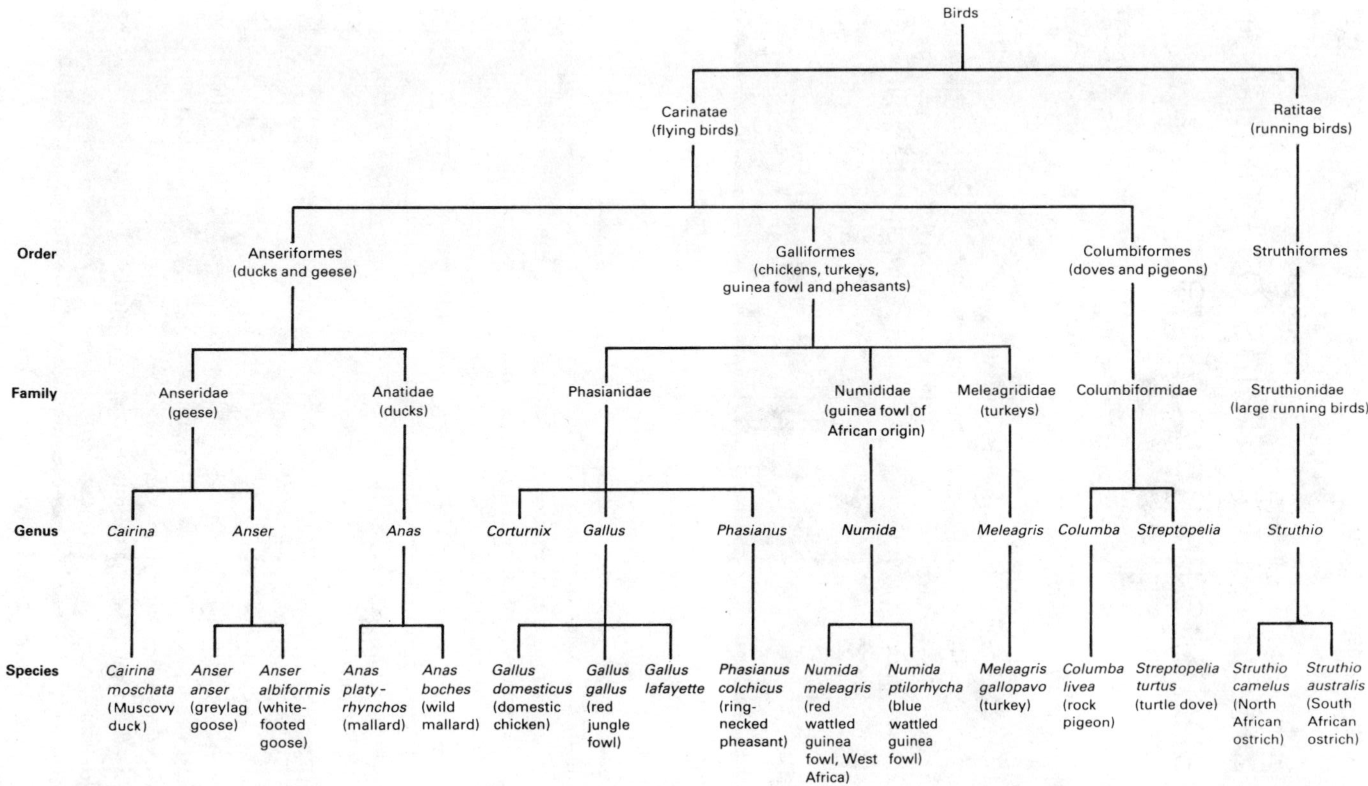

Fig 2.1 *Taxonomic relationships between various domestic fowl exploited by man*

Fig 2.2 *White Leghorn pullet*

Fig 2.3 *280 eggs – the number that can be produced by a White Leghorn or similar breed in one year*

in less intensive systems of production. The cocks are used for meat production and the hens for the production of both eggs and meat. Both are considerably older when slaughtered than are broilers and therefore have more flavour. Older birds are more suitable than broilers for boiling which is a common form of meat preparation in developing countries.

Some better known breeds of *Gallus domesticus* listed by country of origin are as follows:

- **American**
  Rhode Island Red, New Hampshire, Plymouth Rock, Wyandotte (Fig 2.4), Jersey Giant
- **English**
  Orpington, Australorp, Light Sussex (Fig 2.5), Old English Game, Dorking, Indian Game (Fig 2.6)
- **Mediterranean**
  Leghorn (Fig 2.2), Ancona, Minorca
- **Dutch**
  Barnwelder, North Holland Blue
- **Egyptian**
  Fayoumi
- **French**
  Campine (Fig 2.7)

Fig 2.4 *White Wyandotte cockerel*

Fig 2.5 *Light Sussex cockerel*

Fig 2.6 *Indian Game hen*

Fig 2.7 *Campine cockerel*

Dual-purpose birds are particularly suitable for improving scavenging flocks and examples of the use of such breeds in cockerel exchange schemes in developing countries are legion, e.g. the Black Australorp in Malawi, the New Hampshire in Nigeria and the Rhode Island Red and Light Sussex in some countries in Francophone Africa.

The domesticated turkey

The turkey (Fig 2.8) was probably domesticated in the first instance in Mexico, from the Mexican sub-species *Meleagris gallopavo*, and it was used as a domestic fowl by Indian communities in what is now the south-western USA. Turkeys were introduced into Europe in the 1500's,

Fig 2.8 *Norfolk Black turkey*

probably into Spain in the first instance as a result of European colonisation of Central America. The Central American species is small while the species native to the USA is larger and has a characteristic bronze plumage. From Spain turkeys spread throughout Europe. The various European varieties were developed by selection from these imports and these breeds were reintroduced into the USA. The traditional breeds are:

- **Norfolk Black** – medium sized birds, stags 7–8 kg, hens 4–5 kg. Good meat conformation, ninety eggs a year, black quills in skin a disadvantage (Fig 2.8).
- **Mammoth Bronze** – a large bird, stags weigh up to 22 kg, hens up to 18 kg. This strain increases in weight up to 8 kg before it puts on fat. Narrow breast, low egg production (fifty per year).
- **Broad-breasted Bronze** – selected from above for a better breast conformation. Poor fertility unless AI is used. Stags sold 8–9 kg.

In some countries there is a tradition of eating turkeys on special occasions only (Christmas in the UK, Thanksgiving in the USA) but there is absolutely no reason why turkey meat should not be consumed all the year round. For example in the Cote d'Ivoire, the Beltsville White Turkey is produced all the year round to meet the requirements of tourists. Turkey meat is very high in protein and low in fat and therefore has a high feeding value. There are currently about 250 million turkeys in the world and 150 million of these are found in developed countries.

Domestic ducks

The wild mallard (*Anas boscher*) is generally regarded as the ancestor of all domestic breeds of duck, with the exception of the Muscovy (*Cairina moschata*) (Fig 2.9) which is a different species originating from South America. The Muscovy is an extremely good forager and thrives well under free-range conditions. It also has the advantage that it is less fatty than other water-fowl.

Ducks that are kept commercially are kept primarily either for meat production – the Aylesbury and Pekin (Figs 2.10 and 2.11) or egg production – the Khaki Campbell (Fig 2.12).

Of the 500 million ducks kept worldwide, 430 million are found in Asia.

Domestic geese

Domesticated geese (*Anser* spp.) are all thought to be descended from the wild grey goose of Europe and swan goose of Asia. They were probably domesticated at least 3000 years ago. In general the domesticated breeds are much larger than their wild ancestors and they have almost lost their ability to fly. Geese have a major advantage over other classes of poultry

Fig 2.9 *Muscovy duck*

Fig 2.10 *Pekin ducks*

Fig 2.11 *Pekin duck rearing in China*

Fig 2.12 *Khaki Campbell duck*

Fig 2.13 *Flock of scavenger geese*

in that they are exceptionally good grazers and will eat far more grass and herbage. Consequently they are extremely valuable in developing countries where they can be used to graze in places where animals such as goats, sheep or cattle would cause damage to crops. They are used fairly widely in Egypt as scavengers around the villages (Fig 2.13) and historically have been used to weed cotton, although under modern conditions modern spraying programmes may preclude this.

The main breeds of importance for meat production are the Toulouse, Oie des Landes, Embden, Roman, American Buff and Pilgrim. These breeds have descended from the greylag goose whilst the Chinese breeds are thought to have descended from the swan goose.

The Toulouse goose is grey and white. The mature weight of the male is about 13 kg and that of the female 8 kg. It will lay about forty eggs per season.

The Emden is slightly heavier than the Toulouse, white in colour and produces slightly fewer eggs.

The Roman is a fast maturing breed, light in weight (5–6 kg) and capable of producing eighty eggs per year.

The Chinese goose is even more prolific, laying up to 100 eggs a year and it adapts well to hot climates.

Guinea fowl

The guinea fowl is native to West Africa but they are now kept in many parts of the world and are kept on a commercial scale in France and

Fig 2.14 *Guinea fowl*

Italy. Although guinea fowl (*Numida meleagris*) are basically wild birds that are very timid and roost in trees at night, they can be reared intensively (Fig 2.14). They can be kept for egg or meat production. They are light sensitive and will produce ninety to 170 eggs in a year. The eggs have an average weight of 45 g. The shell of their eggs is much thicker than that of eggs produced by chickens and as a consequence they can be transported more easily in trucks along bad roads with less danger of breakages. They can be reared at the same stocking density and under the same environmental conditions, but using a somewhat higher ambient temperature, as broilers and the laying birds kept on the same lighting regime as laying chickens. The young are known as keets. They grow more slowly than chickens (1.5 kg in fourteen weeks) but can be fed on a broiler ration. Their slow growth rate and poor conversion ratio mean that their carcasses have to sell for a higher price than broilers.

Doves and pigeons

Doves and pigeons (*Columba livia*) are used as scavengers in some countries (Fig 2.15). In England in feudal times only the lord of the manor was allowed to own such birds because they would damage crops for many kilometres around the dovecote. They normally lay two eggs in a clutch

Fig 2.15 *A dovecote in the Nile delta (Egypt)*

and the young which hatch in sixteen to seventeen days are fed on 'crop milk' which enables them to grow rapidly.

Pheasants

Pheasants (Fig 2.16) are bred mainly as game birds, i.e. they are bred, released and shot for sport. Some have been reared intensively but they have a very poor food conversion ratio.

Quail

Quail (*Coturnix coturnix*) is native to Asia. It is small but very precocious and can start to lay eggs when six weeks old and may produce as many as 300 in a year. It prefers a temperate climate and therefore is of limited value in the tropics.

Fig 2.16 *Cock pheasant*

Ostriches

The ostrich (*Struthio camelus*) has been farmed in South Africa for over 100 years. Ostriches are kept mainly for their feathers although the eggs and meat are suitable for consumption and their skin makes good quality leather. They are usually fed by rotational grazing on irrigated alfalfa paddocks. A full description of ostrich farming is given by D.J.V.Z. Smit.

Breeding systems and types of inheritance

Historically in the west breeding policies were developed by individual farmers who set their own standards for their stock and selected birds that exhibited what were seen to be desirable characteristics. Birds which showed undesirable characteristics were culled. The selected birds were then bred together in the hope that they would also show these desired characteristics. The success that they achieved depended to a large extent on the characters for which selection was carried out. In many cases selection was for appearance rather than production and many fancy breeds were produced. This type of selection often involved selection for characters controlled by one gene or a few genes. Within this category are included such characteristics as comb shape and colour of plumage. This type of inheritance is termed **qualitative inheritance**.

However, most production characteristics such as body weight, egg production and food conversion are influenced by many genes. The expression of these genes in turn may be influenced by the environment. For example, a laying hen that has been bred to produce a large number of eggs will not be able to produce these eggs unless she is fed an adequate amount of food. This is a classic case of genotype-environment interaction. The inheritance of genes of this latter type is referred to as **quantitative inheritance**.

Sex cells, chromosomes and genes

The carriers of inheritance called **genes** are located on **chromosomes**; these chromosomes are found in all the cells of the bird's body. The numbers of chromosome are distinct for each species but are the same for different breeds of the same species. The numbers of chromosomes found in the body cells of domestic poultry are as follows: chicken 39; turkey 23–39; and duck 39. The most important chromosomes are the ones found in the sex cells because it is these that pass on characteristics to the next generation. When the sex cells, the sperm in the male and the egg in the female, are formed, the number of chromosomes is halved.

If this were not so the number of chromosomes in the cells would double each generation. This means that each offspring receives half its chromosomes from one parent and half from the other. Thus the offspring also receives only half its genes from one parent.

Qualitative inheritance

The transmission of genes and chromosomes from parent to offspring is most clearly seen in qualitative inheritance. In this type of inheritance characteristics are inherited in a simple **Mendelian**[1] fashion and their effects are not influenced by either environment or nutrition. When such characteristics are clearly visible, if appropriate crosses are made, the Mendelian ratios will appear. This type of inheritance is seen in common comb shapes in chickens. There are three comb shapes, rose (R), pea (P) and single (recessive). When a cockerel with either a rose comb or a pea comb is crossed with a hen with a single comb, only chickens with rose or pea combs are produced. In this case the genes for pea and rose are said to be dominant to the gene for single combs. This is true whether the dominant comb form appears in the male or the female. However, if chickens with the two different dominant comb shapes are crossed, a completely new comb shape emerges which is called a walnut comb. This phenomenon can be explained in the following way (Table 2.1)

The above example serves to illustrate qualitative inheritance but is of little practical significance especially since most commercial strains have double recessive single combs (pprr).

Table 2.1 Comb shape and genotype in four breeds of chicken

Breed	Comb shape (phenotype)	Genotype
Wyandotte	Rose	ppRR or ppRr
Indian Game	Pea	PPrr or Pprr
Sussex	Single	pprr
Malay	Walnut	PPRR* or PpRr*

* When the genes for rose combine with those for pea a new comb shape is formed – walnut.

1 Named after the Austrian monk, Gregor Mendel the 'father' of modern genetic theory.

The inheritance of sex

This is a special case of qualitative inheritance since the sex of an individual is determined by the presence or absence of **whole chromosomes**. In the cells of both mammals and birds there are two basic types of chromosome: **sex chromosomes** which determine the sex of the individual, and **autosomes** which determine characters other than sex. For mammals, individuals with two identical sex chromosomes, denoted XX, are female. Male mammals have two dissimilar sex chromosomes, denoted XY. For birds it is the other way round: individuals with two identical sex chromosomes are male, and their sex chromosomes are denoted ZZ, whilst female birds have dissimilar sex chromosomes, denoted ZW.

- In mammals: XY is male
  XX is female
- In birds: ZZ is male
  ZW is female

Sex linked inheritance

One example of qualitative inheritance that does have practical value is the inheritance of certain plumage colours in birds which enables the sex of the chick to be determined at hatching by its colour. The alleles[1] for either gold or silver plumage are located on the Z sex chromosomes, and control sex-linked characters. (The W chromosome in birds is similar to the Y chromosome in mammals and does not carry any genes.) The two alleles of the gene will be located at the same **locus**[1] on the chromosome. The allele for silver is dominant over the allele for gold. Certain well known breeds carry either one or the other of these alleles:

Dominant silver	**Recessive gold**
Light Sussex	Rhode Island Red
White Wyandotte	Brown Leghorn
Ancona	Buff Orpington

1 Certain alleles are found at the same place or locus on chromosomes in a species. Only one of these alleles can be found at a particular locus. Alleles are different forms of a single gene which controls a particular character, colour of plumage in this example.

If a gold cockerel of the Rhode Island Red breed is mated with a silver hen of the Light Sussex breed the resulting chicks will be either silver or gold. The cockerels will always be silver and the hens gold thus the colours have crossed over between sexes in the offspring as compared with the parents.

Parents	Gold cockerel	Silver hen
Offspring	Silver cockerel	Gold hen

Why should the changeover occur? This is quite simple to understand if the genetic make-up of colour inheritance is examined. A gold cockerel has two Z chromosomes each carrying the recessive gene for gold (s) whilst the silver hen has only one Z chromosome carrying the dominant silver gene (S).
When the silver cockerel and the gold hen are mated, approximately half the chicks of each sex are silver and the other half are gold.

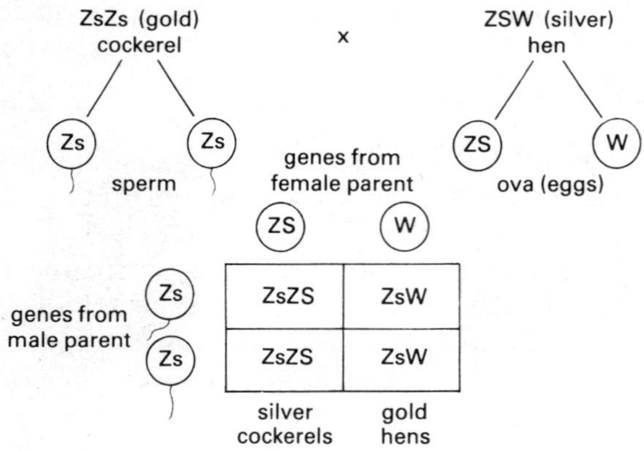

Thus it is possible to determine the sex of the chick at hatching without resorting to vent sexing. However the method will work only if the dominant gene is introduced from the female. If it is introduced from the male either all the offspring will be silver (if the male is **homozygous**, like alleles, for this gene) or half the offspring of each sex will be silver (if the male is **heterozygous**, unlike alleles).

The main problem with the use of the above type of **sex linkage** is that it works only with certain types of cross (i.e. male with recessive alleles for colour and female with dominant allele for colour). An attempt to overcome this problem was by the use of the barring breed and the associated development of auto-sexing breeds of poultry.

Auto-sexing breeds

An allele for **barring** is located on the Z chromosome of certain individuals. This allele reduces pigmentation in the plumage. This is more acute in males which contain two barring alleles (BB) than females which contain only one (BW). The advantage of using this allele is that once it is introduced into a breed from which all the barring alleles of the gene had been removed from the population by selection, the chicks of each sex are distinguishable at hatch. Using this method; the ability to identify sex does not depend on crosses between different breeds, and is known as auto-sexing.

In the 1930's auto-sexing breeds (Figs 2.17 and 2.18) were developed. These included Dorbars (from Dorking chickens) and Legbars (from Brown Leghorns). Auto-sexing breeds never developed commercially for two main reasons. Firstly, in the 1930's vent sexing by the Japanese method was developed, and this reduced the need for auto-sexing. Secondly, the White Leghorn became the most important parent of most hybrid strains of egg producing birds, and auto-sexing does not work with White Leghorns. One auto-sexing breed that is frequently used in Senegal, Upper Volta and Cote d'Ivoire is the cross between the Rhode Island Red and the barred Plymouth Rock.

Fig 2.17 *Chicks of an auto-sexing breed, male (left) contains two barring alleles and consequently the colour dilution (white spots on head) is more marked than in the female (right)* (Chris McKinnell)

Kk alleles in chickens

The rate of feathering of chicks after hatching is determined by the alleles K and k. This has been rather more useful for sexing chicks than the barring allele. The recessive allele k is responsible for rapid feathering whilst the dominant K allele is responsible for slow feathering (Fig 2.19). A cross between rapidly feathering kk males and slow feathering KW

Fig 2.18 *Cockerel (left) and hen of Campine breed, an auto-sexing barred breed*

Fig 2.19 *Differences between slow and rapid feathering chicks* (Chris McKinnell)

females results in slow feathering Kk males and fast feathering kW females (Fig 2.20). The chicks can be distinguished by the rate of development of their first wing feathers. These alleles can be introduced with any breed but they suffer the disadvantage of working only when the specific cross referred to above is made.

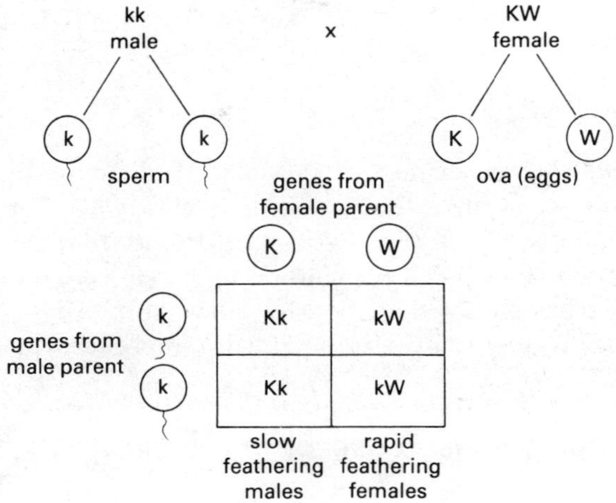

Fig 2.20 *Effect of crossing rapid feathering males with slow feathering females*

Sex limited inheritance

Although the genes for certain characteristics are found in all individuals, both males and females, they are expressed in only one sex. An example in poultry is the gene for egg production.

Quantitative inheritance

For most characters of economic importance in farm poultry it is not possible to classify animals in one or another discrete class. For example, egg production by laying birds may range from 100 eggs per annum to 300 eggs per annum. Within this range there is continuous variation. Characters such as egg production are described as quantitative and the science which has been developed to explain their inheritance is called quantitative genetics. The basis of this science is that quantitative characters are influenced by many genes and normally also by environmental influences. For these reasons, knowledge of the phenotype of the bird does not give a direct indication of its genetic make-up as would be the case with qualitative characters. Similarly it is not possible to make predictions of expected frequencies of different genotypes in the population. Consequently, when thinking of the inheritance of these quantitative characters one has to think not in terms of individual animals and particular mating pairs. Instead one thinks in terms of populations of animals and the frequency of occurence of various alleles within the populations.

Population genetics

A **population** may be defined as the total of all the individuals in a breed. This could mean within a breeder's closed flock or all the individuals within the *Gallus domesticus* species. **Population genetics** is the study of genetic make-up of a population. Research work on the genetics of populations has shown that the Mendelian laws of inheritance that apply when individual crosses are made also apply to populations.

Forces changing allele frequency in a population

Mutation pressure

Up to this point we have assumed that genes are unchanging parts of chromosomes that determine inheritance. However genes do change occasionally to give out new instructions. If they do this they are said to have **mutated**. The rate of mutation of genes can be increased artificially by irradiation of the sex cells or by treating them with certain chemicals. Mutation of, for example, allele 'A' to allele 'a' increases the frequency of 'a' in all subsequent generations and decreases the frequency of 'A'.

Differential migration

The immigration of new alleles of a gene into a population can be entirely controlled by the breeder. This is achieved largely by crossing an existing population with another, thus introducing new alleles and therefore changing relative allele frequency. A cockerel exchange policy is a way of bringing about the immigration of alleles into a population. The emigration or loss of alleles from the population may occur due to inbreeding and will occur if selection takes place in a population.

Genetic drift

The frequency of alleles in a gene pool should remain constant with time unless mutation, selection or differential migration affect their frequency.

However constancy exists only in infinitely large populations. In reality populations consist of a finite number of individuals. Under these circumstances allele frequency can change because more alleles of one gene are lost from a population by chance than other alleles of the gene. Such changes will take place more quickly in small populations than in large ones.

Selection

Selection is the process by which certain individuals are preferred to others to become parents of the next generation. Selection is of two types, that which occurs naturally and that due to the efforts of man. Selection is the most important method of changing allele frequency in breeding programmes. It is therefore important to know what factors affect selection efficiency.

Factors which determine selection efficiency

Selection is one of the tools open to a poultry breeder when he wishes to change the allele frequency in a population. In selection programmes within an existing population the poultry breeders have but two decisions to make.

- To decide which birds should be used as the parent of the next generation.
- To decide which individual cock should be mated with which individual hen.

Therefore it should in theory be a straightforward task to design a simple and effective programme to improve a population of birds in the way the breeder desires. However, there are a number of constraints to this process.

Ability of the breeder to identify superior birds

In order to achieve change in the allele frequency in a population, the breeder must have definite objectives, e.g. rate of egg production; growth rate; egg weight. Once these objectives have been identified, improvement can be achieved, but genetic superiority may be masked by a number of factors. Identification of superior birds is particularly difficult if the trait is sex linked, as in the case of egg production. How can you select cocks for traits associated with egg production? The answer of course is by using information derived from the egg laying ability of their female relatives.

In order to be able to identify superior birds the breeder must keep adequate records on female performance. Comparisons between individual birds should be made after appropriate corrections have been made for age, sex and other relevant factors.

Variation present in the populations at the start of the selection programme

The more genetic variation that exists in a population at the start of a selection programme the more rapid is likely to be the progress towards defined objectives. In a previously unselected population there is likely to be a large degree of variation in, for example, indigenous flocks of poultry. In a population that has previously been inbred or heavily selected the amount of genetic variation may be very low indeed. Crossing two inbred populations before the start of the selective process will increase the genetic variations within it and therefore make selection more efficient.

Amount of selection pressure applied

The selection pressure applied to a population depends upon the difference between the mean performance of the birds selected for breeding and the mean of their generation. The differences between the flock average and that of individuals used for breeding is known as the **selection differential** (Fig 2.21).

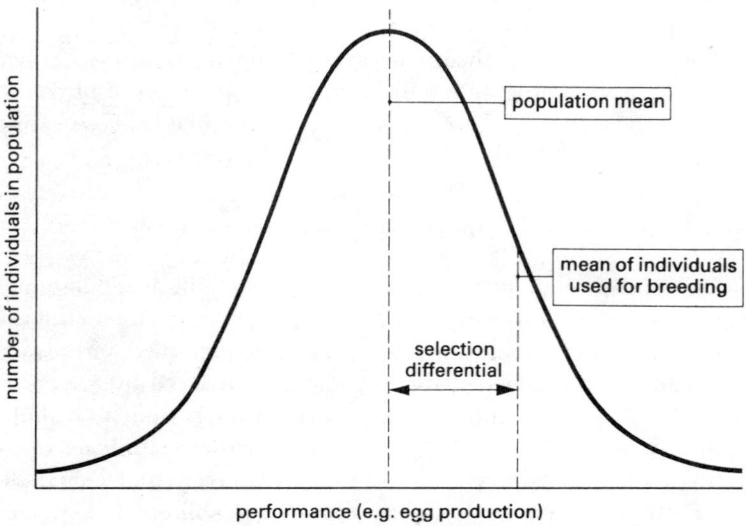

Fig 2.21 *Performance of flock for traits under consideration (e.g. egg production)*

In general terms the larger the selection differential the more is the progress that can be expected in the selection process. A number of factors may affect the size of the selection differential, amongst these are the number of birds that need to be kept for replacement purposes. More birds will need to be saved and therefore the selection differential will be less in populations that have a high mortality rate and/or are increasing in size. The selection differential for males is always greater than for females because few of them need to be saved for breeding purposes. In poultry flocks, which have much higher reproduction rates than mammals, the selection pressure can be much higher than for mammals.

Heritability of traits

Heritability is normally referred to as a heritability estimate because precise measures are impossible to determine. Heritability is referred to by the symbol 'h^2'. The heritability estimate for a trait in the population is based on the amount of the variation for that trait in the population that is due to genetic variability. Let us assume that the heritability estimate for growth in broiler chickens is 50% and that the average daily growth is 40 g. This does not mean that 20 g of growth is due to inheritance and 20 g due to the environment. What it does mean is that of the differences between individuals in the flock 50% are due to heredity and 50% are due to environmental effects. Thus an estimate of heritability is made by dividing genetic variation by the total variation amongst individuals.

$$\text{Heritability } (h^2) = \frac{\text{genetic variation}}{\text{genetic variation} + \text{environmental variation}}$$

The term heritability may be used either in the wide or narrow sense. In the **narrow sense** heritability estimates indicate mostly additive gene action. This can be used to estimate the amount of genetic progress that is likely to be achieved in the next generation when superior individuals are selected as parents. Heritability in the **broad sense** includes all the effects of genetic variability and in addition to additive gene action includes effects of dominance and epistasis. This type of estimate will not give a reliable estimate of progress to be made by selection. Estimates made from selection experiments give the best estimate of heritability in the narrow sense followed by estimates made from phenotypic similarity amongst relatives. Thus it should be clear that the amount of progress that can be made in the selection process is affected by the heritability of the trait. Selection for a trait that has a low heritability, such as egg production in chickens (5–10%) will result in little progress being made. On the other hand, selection for a trait that is highly heritable such as body weight in broilers (60%), should result in rapid progress being made in the mean performance of the flock. If the heritability of a trait is high a large proportion of the superiority of the parent will be retained in the offspring, as is shown in the following example.

Rate of gain in broilers:

Heritability of rate of gain	60% or 0.6
Flock average rate of gain	35 g a day
Rate of gain of cockerels saved for breeding	50 g a day
Rate of gain of hens saved for breeding	40 g a day

$$\begin{aligned} \text{Expected genetic gain} &= h^2 \times \text{selection differential} \\ &= 0.6 \times \left[\frac{(50 + 40)}{2} - 35\right] \text{ g} \\ &= 0.6 \times (45 - 35) \text{ g} \\ &= 0.6 \times 10 \text{ g} \\ &= 6 \text{ g} \end{aligned}$$

$\therefore$ average expected gain of offspring = 35 g + 6 g = 41 g

If the heritability estimates for traits in the narrow sense are low this indicates that additive gene action is probably small and that non-additive gene action such as dominance and epistasis is probably large. With such traits cross-breeding is likely to be far more effective than selection.

Generation interval

This may be defined as the average age of parents when their chicks are hatched. In the case of chickens and ducks the generation interval is about eight months and it is somewhat longer for turkeys. The shorter the generation interval the faster the rate of annual genetic change.

Genetic correlation amongst traits

Quantitative traits can be influenced by many genes each having only a small effect. When two traits are affected by the same genes or linked genes, they are said to be genetically correlated. This phenomenon, which is known as **pleiotropy** is probably the major cause of genetic correlation although it is possible that **linkage** (the close association of genes on the same chromosome) may have similar effect.

If the genetic correlation between two traits is high, selection for one will ensure selection for the other. Thus if the selection is made for rate of gain in broilers, selection will also be achieved for efficiency of food conversion. However, if two desirable results are negatively correlated these can pose considerable problems for the poultry breeder. The classic example is the negative correlation between egg production and body weight increase in broiler breeders. A similar problem is posed by the negative correlation that occurs between egg production and egg weight when these two traits are selected for in laying stock.

Number of traits in the breeding programme

The more traits that are selected for in a breeding programme the slower the progress made for any one trait. The rate of progress for a number of traits is proportional to $\frac{1}{\sqrt{n}}$ where 'n' is the number of traits and '1' is the rate of progress achieved when selecting for one trait. The message is to select for as few traits as possible, do not select for 'fancy traits', e.g. comb shape, or characteristics with very low heritability or characters that can best be modified in other ways, e.g. yolk colour.

Methods of selecting for several traits at the same time

In poultry production, in order to produce a bird of commercial value, it is necessary to select for a number of traits at the same time. This is not an easy process because the traits may not be of equal economical value nor will they always be genetically independent. Three methods of selection for multiple traits are generally recognised: the tandem method, independent culling method, and the selection index method.

The tandem method

In this technique traits are selected for one at a time, for example, the breeder could select first of all for the egg production ability in one generation and then for mean egg weight in the second generation. The techniques will work if the traits selected for are positively correlated, not as in the above case, negatively correlated. If the traits are negatively correlated the progress made in the first generation will be partially reversed in the second. The tandem method is the least efficient of the three methods available.

Independent culling method

With this method the breeder lays down a minimum standard for several traits. Each bird to be used for breeding must achieve all of these standards, if not, it will be culled. This technique is commonly used in traditional and pedigree breeding. For example, the following standards might be laid down for a broiler hen (Table 2.2).

Table 2.2 Criteria that could be used for selecting egg laying hens by the independent culling method

	Weight at 8 weeks of age (kg)	Age at 1st egg (days)	Mean weight of first 20 eggs (g)	Eggs produced during first 50 days of laying
Standard	1.8	140	50	25
Bird 1	1.7	135	55	30
Bird 2	1.9	133	48	32

With the given 'standard', 'bird 1' would be rejected for breeding because it was too light at eight weeks of age and 'bird 2' would be rejected because it produced eggs that were too small. This would result in those birds being rejected even though they are above the required standard in all the other traits and therefore would have probably been desirable parents of the next generation. Thus although the independent culling method is superior to the tandem method in that it allows for selection of more than one trait at a time, it is seriously flawed as a selection technique. In addition to the disadvantage cited above this technique makes no allowance for the economic worth of the trait nor its heritability.

The selection index

In this method a numerical value is given to each potential parent for each of the traits regarded as important. These numerical values are summarised and a total score arrived at for each bird. The birds with the highest numerical values are the ones saved for breeding purposes. The influence that each trait has on the final index is determined by how much importance that trait is given in relation to other traits. The amount of importance given depends on its relative economic value, upon its heritability and the genetic correlation between traits.

In poultry breeding the calculation of a selection index is a very complex process. It is usually impracticable, except in very large establishments for an individual breeder to calculate appropriate heritability and correlation figures for his own flock. However, in unselected populations of poultry average published figures of heritability estimates can be used initially although these may become unreliable once selection proceeds. Selection may reduce heritability estimates and genetic correlation may change greatly with time. The large-scale breeder will therefore automatically verify these at predetermined intervals.

The problem of weight for economic worth is even more difficult because values may change from year to year or from market to market.

For example, premiums paid for large eggs may mean that egg size should get a high economic weighting. An even more complex problem of deciding economic value occurs when the breeder is selecting for traits, some of which are important to him directly and some of which are important to his clients. For example, the grower of broiler chickens is interested in growth rate, food conversion and carcass quality. He is not interested in fertility and hatchability. To be a financial success the breeder has to produce enough chicks for sale, but to obtain repeat orders his stock must satisfy the grower. It is a very complex problem to solve, especially when some of the traits are negatively correlated.

Breeding systems

Traditional breeding systems

When poultry breeding was in the hands of the traditional breeders, selection was carried out on the basis of the appearance of the individuals. This type of breeding is still carried out with the 'fancy breeds' of poultry, where breeding is for show points. With such a system the best individuals are selected, and these are crossed with other individuals carrying traits deemed to be good by the breeder. To be successful, this system depends on the breeders being able to identify 'good' individuals, and it also depends on whether or not the characters for which selection is being made are 'heritable' to any marked extent. This type of system is likely to be most successful with characters that are easy to see and measure such as shape of comb or colour of plumage, and with those characters that are highly heritable like body weight. This method is likely to be far less successful with characters that have a low heritability and can only be identified in one sex, such as rate of egg production.

Inbreeding

Inbreeding is a mating system in which the progeny are produced by parents that are more closely related than the average of the population from which they came. The main problem with inbreeding is that it tends to unmask numerous genetic defects in the populations which result in a decline in vigour and performance.

Cross-breeding

The effects of cross-breeding are the exact opposite to those of inbreeding. The reasons for cross-breeding are to increase genetic variation in the population and to encourage the exhibition of **hybrid vigour**. This latter reason is particularly important with characters of a low heritability such as egg production.

The heritability of hen housed average egg production (page 40) by laying birds can be as low as 5%. Of the total recorded variability in egg production between hens, 5% results from the effects of additive genes, the remaining 95% results from a combination of environmental and non-additive gene effects such as dominance and epistasis.

With this type of character, selection is likely to be ineffective and cross-breeding is likely to be more effective because hybrid vigour is likely to occur. This is also true with other characteristics of a low heritability such as fertility, hatchability and survival. Modern systems of breeding aim to exploit heterosis (hybrid vigour) to the full. Therefore parent stocks are bred for their ability to combine genetically with other stocks. Parent stocks should be inbred to obtain a high level of homozygosity and then the inbred lines crossed. A coefficient of inbreeding of at least 50% is needed (three generations of full sib mating) in order to secure reasonable homozygosity and repeatability of performance. Such degrees of inbreeding are possible with chickens, because of their high rates of reproduction and low unit costs. Even so the wastage rate can be very high. Wallace, who was probably the first to work with poultry breeding on a large scale (in the 1920's, 1930's and 1940's), found that inbreeding uncovered a good deal of genetic 'trash' and as a result 96% of the inbred lines that he started died out because of infertility or they proved unsatisfactory. However the remaining 4% showed the advantages of this system of breeding, namely a uniform product and the possibility of heterosis when crossed with other lines. However, crossing is an uncertain business and not all crosses are successful and there is no way of knowing before a cross is made whether or not it will be successful. If two lines do cross successfully and produce productive offspring they are said to '**nick**'[1]. If the cross is between two strains of the same breed it is called an incross and if between two strains of different breeds an incross-breed. This type of breeding is produced by large breeders in America and Europe but it is too expensive for breeders in developing countries who will normally import hybrid birds from these countries.

Types of cross-breeding

Pre-selected strain crossing

This system of breeding, which exploits hybrid vigour, can be practised by fairly small-scale breeders. Two strains A and B are established. Individuals in flock A may nick with individuals in flock B for the characteristics desired, such as egg production. For the system to be

1 **Nicking** is a term used by poultry breeders to indicate that two strains of breeds when crossed will produce offspring that are superior to either parent.

successful each flock should contain a high proportion of complementary individuals and the individual flocks should have been closed for a sufficient period of time for individual birds to be sufficiently homozygous in their gene make-up to ensure repeatability from season to season. The two flocks should be tested one against the other until two complementary strains are exposed. Selected individuals can then be employed for strain reproduction by traditional methods of breeding. The balance of each flock is cross-bred in the hope that each will supply sufficient individuals which nick with each other in the other strain of bird.

Reciprocal recurrent selection

Birds of two strains are reciprocally crossed (A males with B females and vice versa) and the progeny are tested for performance. The sires and dams which produce the best cross-bred progeny are used as parents within their own strain to produce the next generation of parents. The cross-bred progeny are sold for commercial use. The balance of the flocks not being tested are cross-mated for the mass production of commercial chicks. Thus reciprocal recurrent selection is based on one year for the selection of additive gene values and in the second year on heterosis.

Breeding for egg producing hens

The principal objective of the breeder is to produce a strain of poultry that will lay the maximum number of eggs within 500 days from hatch whilst consuming the minimum amount of food. Other factors have also to be taken into account such as egg weight, body weight, feed consumption, viability and egg quality. These are the factors that concern the producers of table eggs but the poultry breeder must also consider fertility and hatchability. Since so many traits are involved, a selection index is essential and because many of the characters have a low heritability, a cross-breeding programme is essential.

Egg production

The criterion used is the number of eggs produced within a certain period of the year. The normal period is 500 days from hatch as this measure takes into account the age at which the bird lays its first egg and gives a better score to the individual that matures early. Egg production can be calculated in three ways.

(a) Survivor egg production – that is the number of eggs produced by birds that survive for 500 days.

(b) Hen day production – that is the production of all eggs produced by the flock divided by the product of the number of days and the number of birds alive on each of these days.

(c) Hen housed production – that is the number of eggs produced divided by the number of birds housed multiplied by the number of days that the birds were actually in lay.

The latter is probably the best measure since it not only takes into account the age of maturity of the birds but also their viability. Flocks that suffer large mortality rates during the laying season will obviously appear to have a low productivity rate when measured in terms of hen housed production percentages.

The most important factor limiting the rate of egg production is probably shell formation. Hens produce eggs on average every twenty-five to twenty-six hours (range twenty-four to thirty hours). This means that a bird producing an egg every twenty-five hours will lay the egg one hour later each day. Chickens do not normally lay eggs in the afternoon and therefore once the egg laying is delayed beyond midday, **oviposition** (egg laying) is delayed until the following day. This procedure results in eggs being laid in clutches. Some birds produce an egg every twenty-four hours and therefore could, at least in theory, lay 365 eggs a year. Such birds are however very rare.

The inheritance of hen housed production is very low (5–10%). This is partly because the figures include allowances for viability, broodiness and moulting. The figure for survival production is much higher, possibly as high as 30%. Broodiness is a natural, highly heritable characteristic and because of its high heritability it is easily selected against and as a result is rarely seen in modern high-producing flocks.

Egg weight

Egg weight is a fairly high heritable characteristic which is unfortunately negatively correlated with egg production. As a result of this correlation, selection for high rates of egg production generally result in strains of poultry which lay small eggs. By the use of selection indices, breeders have been able to overcome this. Mean egg weight in chickens increases by about one gram a week for six to seven months and then levels off, so selection should be made between birds at similar stages of lay.

Viability

A breeder wishes his birds to be able to resist a whole range of diseases. Although it has not been shown that such non-specific resistance exists, it is certainly true that consistent differences in viability can be found between strains and crosses even when the comparison is made on several farms. However, the heritability of viability is very low (5–10%) and the best way of selecting for it is to use estimates of productivity based on hen housed averages.

Body weight

In developed countries the tendency has been to select for strains that have small body weights since this ensures a lower maintenance requirement and lower food requirements and therefore more efficient production of eggs. This approach can be justified because in developed countries there are always well-developed meat industries and therefore the market for old hen carcasses is very small indeed. The same consideration does not apply in many developing countries where old hens are preferred to broilers and therefore the hen carcasses may have considerable value. This fact linked to the one that egg weight is positively correlated with body weight may mean that somewhat heavier laying birds are more desirable in developing countries than in developed ones.

The heritability of body weight at a given age is fairly high, 30–60%. A specified age for selection is important because birds do put on weight throughout the laying cycle. The body weight should be measured as soon as possible to reduce selection interval, but not before the birds are twenty-two weeks of age because until then birds are growing rapidly due to sexual development. Probably the best time to measure body weight is when the birds are about twenty-five weeks old.

Egg quality

Egg quality is important in a discerning market. Egg shape, shell quality and colour, albumen and yolk quality are all important. It is obviously less important in markets where eggs are a scarce commodity and when the price of eggs is the most important consideration.

Egg shape

Egg shape is important when eggs are packed in uniform trays for transportation. Abnormally shaped eggs become broken in the handling process because they do not easily fit into these trays. The ideal shape is defined by the relationship between the length and the breadth of the egg. A breadth/length × 100 index of 75 is regarded as the most satisfactory. Egg shape is of less importance when eggs are not packed in specialised containers.

Shell quality

Shell quality, which is related to shell thickness, is a very important characteristic. Eggs with thin shells crack very easily in transit. This can be a serious problem in tropical countries where high temperatures cause laying birds to lay thinner shelled eggs, particularly in the hot season of the year. Shell thickness can be measured with a micrometer or non-destructively by specific gravity. Although other characteristics such as porosity of the shell, membrane thickness, mineral content of the shell and the thickness of the protein matrix all affect shell strength, shell thickness is by far the most important. Strength of shell decreases much more rapidly than thickness. Few eggs with a shell thickness of 35 micrometres (Fig 2.22) break in transit whereas all eggs of 25 micrometres or less break before or during the handling process. The heritability of shell thickness is about 30%.

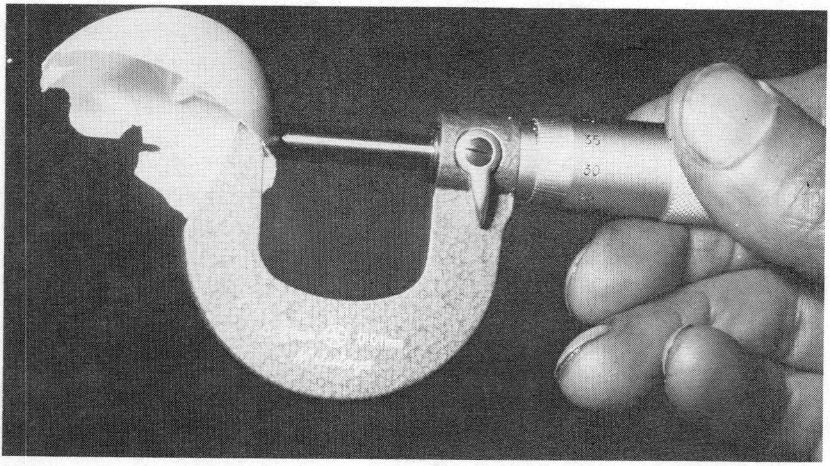

Fig 2.22 *Measuring thickness of the shell of an egg with a micrometer — this shell is about 33 μm thick*

Shell colour

In western countries, such as the UK, shell colour is regarded as important and the myth has developed that brown eggs have a higher nutritional quality than white eggs. However, this is not so and the colour of the eggs is due to a surface covering of the pigment oöporphyrin.

Interior quality

Albumen quality

It is possible to select for albumen quality, i.e. eggs with a greater percentage of thick white. However, other factors such as egg storage and flock health are probably more important factors influencing this characteristic.

Blood and meat spots

Blood and meat spots are of separate origin; both appear on the border between the yolk and albumen. Blood spots are caused by haemorrhaging of vessels in the ovary. Meat spots originate from lesions in the mucosa of the oviduct. The occurrence of both seems to be fairly highly heritable and values of between 25% and 50% for heritability estimates have been obtained.

Yolk characteristics

The colour of the yolk is important to the consumer and is greatly influenced by the food fed to the chicken. There is no point in selecting for yolk colour since all yolks will reach the required level of pigmentation if the correct pigments are fed to laying birds.

The commercial breeding of laying birds

The theory behind this procedure was discussed earlier but to recapitulate, the aim is to breed and select inbred lines and then cross these lines to achieve heterosis. Some idea of the scale of operations in commercial poultry breeding can be seen from the following diagram derived from data for a firm that had a market for ten million layers of one strain (Fig 2.23).

Heterosis occurs when strains A and B are crossed and when strains C and D are crossed. No further heterosis is observed when AB is crossed

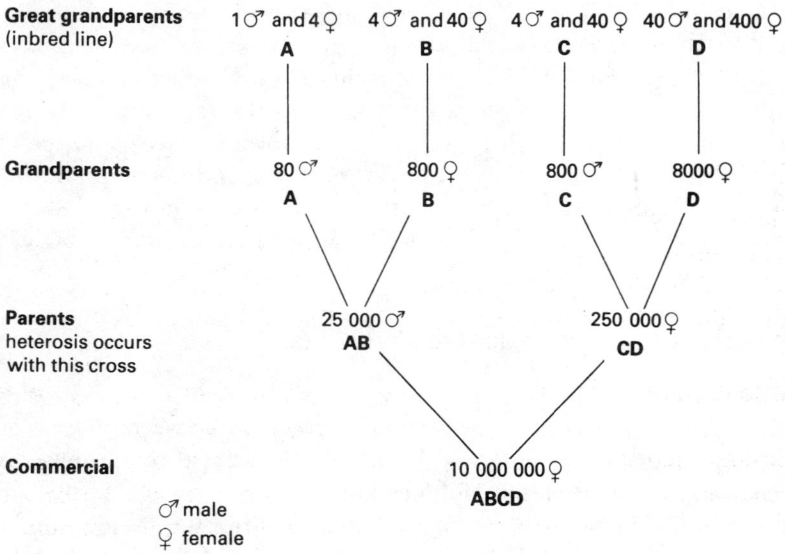

Fig 2.23 *A breeding programme to produce commercial egg laying hens*

with CD. This last cross is necessary however to build up numbers sufficiently to make the number of chicks produced a commercial proposition. Although heterosis is important in poultry breeding, the base for all breeding work is selection. Its aim is to increase favourable gene effects especially additive ones. Within a selection programme the main problem is the estimation of the breeding value of a single bird. Individual records give valuable information in working with highly heritable traits only. Generally the estimation of breeding value is based on a combination of additive gene effects and response to cross-breeding.

A very important point complicating the selection work is the combining ability of the parental strains which is partly due to heterosis as mentioned above. Positive heterosis effects are mainly to be expected in characters of low heritability. Therefore parental strains must be bred for their ability to cross successfully with other strains (combining ability). Theoretically probably the most effective way of breeding for combining ability is reciprocal recurrent selection where males of strain A are progeny tested against selected females in strain B and vice versa (p. 39). This is a good method of developing cross-breds whose superiority is due to specific combining ability. The disadvantage of this system is that it doubles the normal generation intervals.

Breeding for meat production

Breeding for meat production is somewhat more complicated than breeding for egg production because several conflicting requirements have to be taken into consideration. The constraints on the farmer who produces broilers are different from those of the breeder who is producing broiler hatching eggs. The broiler grower wants stock that will reach a high body weight in the shortest possible time on the minimum amount of food. He also requires birds that have the correct body conformation, will feather rapidly, and have a low level of mortality. The producer of the eggs that produce the broilers is more interested in egg production characteristics because he wants to produce the maximum numbers of chickens for sale.

Egg production from broiler breeders

Birds that grow rapidly in the early weeks also tend to be overweight when adult. There is a strong negative correlation between body weight and egg production. Consequently the breeder has to compromise and accept adult hens that will lay only 140 eggs per year and at the same time be difficult to manage. As with laying birds the best results are obtained by using a selection programme followed by cross-breeding. Different breeds are used for the male and female strains. The White

Cornish (white variety of Indian Game) is used for the male line, perhaps crossed with White Rock. The White Cornish tends to grow faster than the White Rock but lays fewer eggs. Also more selection pressure for growth rate can be put onto the male side because egg production is less important. On the female side the White Rock or White Rock type of bird has now largely replaced the Light Sussex or New Hampshire. To increase egg production for the female lines White Rocks may be crossed with New Hampshire or Light Sussex.

The type of breeding programme used could be one of the following:

table quality strain male (Cornish) × egg production strain (New Hampshire)

↓

broiler chick

However, a more likely programme would be a three- or four-way cross as follows:

Cornish male × White Rock → cross-bred male

White Rock male × New Hampshire → cross-bred female

cross-bred male × cross-bred female

↓

broiler chick

Intense selection would be applied in the male strains for meat quality, food conversion and food efficiency. Less intense selection for these characters would be applied on the female side. The use of cross-bred females will ensure hybrid vigour, maximum egg production, viability and hatchability from these cross-bred birds.

Growth rate and food efficiency

In an unselected population the heritability of growth rate in chickens is very high. Heritability estimates of between 40% and 80% have been obtained. However, continued selection for growth rate results in a reduction in genetic variation and as a result, heritability of traits may become very low. This means that these strains may not respond to further selection for growth rate. The 'plateauing' of strains may be overcome if two parental strains are crossed and then selection is made from within the cross-bred individuals. However, selection has been very successful and it is not unusual for male broilers to weigh 2 kg liveweight at six weeks of age. Females will tend to grow at a rather slower rate. The fact that females grow more slowly than males is not necessarily a disadvantage because the consumer demand is for a range of broiler carcasses of different weights.

Efficiency of food utilisation of broilers is very closely correlated with growth rate. Since food intake is difficult to measure it is better to estimate food conversion from growth rates.

Carcass quality

What is considered to be a good quality carcass depends on the consumer. Broilers which are regarded as high quality in western countries are regarded as tasteless and insufficiently chewy in many African countries. Thus carcass quality is difficult to define objectively. However, if the birds are to be marketed in supermarkets, the breast muscles should be wide and deep.

Feather and skin characteristics

White feathers are preferred by producers to coloured ones because they result in a 'cleaner' looking carcass. However in countries where rainfall is heavy and the mud is red, keepers prefer birds with coloured feathers such as Rhode Island Red. A good feather cover is required since this reduces the incidence of heat loss and the presence of breast blisters. Both of these characteristics are heritable.

Most modern strains of broilers will produce yellow fat since they have been selected for the American market where a yellow colour is desirable. In other markets where yellow fat is undesirable the required fat colour can be achieved by avoiding diets containing carotene and carotenoid pigments.

Improving indigenous birds by cross-breeding

The cross-breeding of indigenous poultry with exotic birds is one way of increasing the productivity of local stock. Indigenous birds tend to have a low productivity and do not come into lay until six to ten months of age and when they do lay, produce very few eggs, which are mainly of a small size. Trials in developing countries, including Nepal, have shown that crossing indigenous birds with exotics of traditional breeds such as New Hampshire and Black Australorp can result in early maturity of the flocks and a greater production of eggs of a larger average size. Such a genetic change can be achieved by a policy of cockerel exchange. For this purpose poultry owners are given exotic cockerels in exchange for cockerels of the local breeds. There has been some success with such schemes. For example this method of improvement has been well used in West Africa since 1950 with very good results provided the nutrition of the birds is adequate.

3 The effect of the tropical environment on poultry and possibilities for modification

Introduction

Various factors in the bird's environment affect its well being and its levels of productivity. In tropical areas the relative importance of these vary from area to area. For example, humidity in the air is of more importance near the equator in the rain forest areas whereas, high temperatures are important in the seasonally dry areas away from the equator and very important in hot desert areas. Similarly the effect of day-length on laying birds becomes more pronounced the further they are kept from the equator. It should however be emphasized that because under 'modern' systems of management poultry are normally intensively housed and therefore live in a modified micro-environment. Of the microclimatic factors the two which concern us most are effective temperature and day-length.

Effective temperature

Almost all the energy on the earth originates from solar radiation. The total amount of radiation received on any part of the earth depends on two factors; the duration of the radiation and its intensity. Both of these factors are controlled by the rotation of the earth and its movement around the sun. The higher the elevation of the sun, the more intense the radiation because its rays are spread over a smaller surface area and because the rays have to pass through less atmosphere and are therefore less liable to diffraction by dust particles in the atmosphere. Thus the sun is hotter in equatorial regions than in temperate regions, although this effect is modified in heavy rainfall areas by the presence of a thick cloud cover. Thus the places of highest temperature are found not on the equator but in the summer in latitudes above 14°N and below 14°S where a lighter cloud cover is associated with longer day-length.

When considering the effect of tropical temperature on poultry it is important not just to take into account average annual temperature but also seasonal and diurnal variation and also the modifying effects of air humidity and wind speed. For example in low-lying wet areas the mean group temperature is not particularly high, for example in Singapore it is in the region of 26–27°C, but this temperature is associated with a high average relative humidity. Also the annual and diurnal range of temperature is only 2–3°C and the wind speed in these areas is also normally very low. On the other hand in dry desert regions, although the annual mean temperature may be similar, the range of temperature may be from lower than 0°C to over 45°C in the hottest month. These areas will have a low relative humidity (except during the rainy season) and the mean wind speed will be considerably higher than in the humid zone.

The effective temperature is also affected by altitude. This is because the heating of the atmosphere is an indirect process; solar radiation heats the earth's surface which in turn heats the atmosphere. Thus the temperature of the air decreases with elevation. The rate is variable but is normally around 0.5°C per 100 metres.

The effective temperature is affected by humidity and air movement (Fig 3.1). Humidity is important at high temperatures because it affects the ability of birds to lose heat by evaporation. Thus high temperatures associated with high humidities are more stressful than high temperatures associated with low humidities.

Day-length

All breeding poultry are influenced by changes in day-length but the effect on the female sex is the most important commercially. This effect has been extensively quantified for the laying hen and also to some extent for the turkey and the duck. Egg production is stimulated by increases in day-length, whilst reduction in day-length can lead to a cessation of production and to the birds moulting. Thus in modern poultry keeping, birds are exposed to lighting environments modified to stimulate egg production.

Under natural lighting conditions day-length varies with time of year and latitude. It is only at the equinoxes (23 March and 22 September) that the days and nights are of equal length everywhere. At the equator the effective day-length is always a little over twelve hours, but with increasing latitude there is an increasing difference between the longest day in summer and the shortest day in winter.

Controlling day-length for laying hens in temperate countries presents little problem because the birds are normally kept in windowless houses. However in the tropics, open-sided houses are the norm and therefore supplementary lighting rather than complete control is the most normal solution. For example, in the humid zones where day-length changes little throughout the year, two to three hours of artificial light is all that is needed for laying birds.

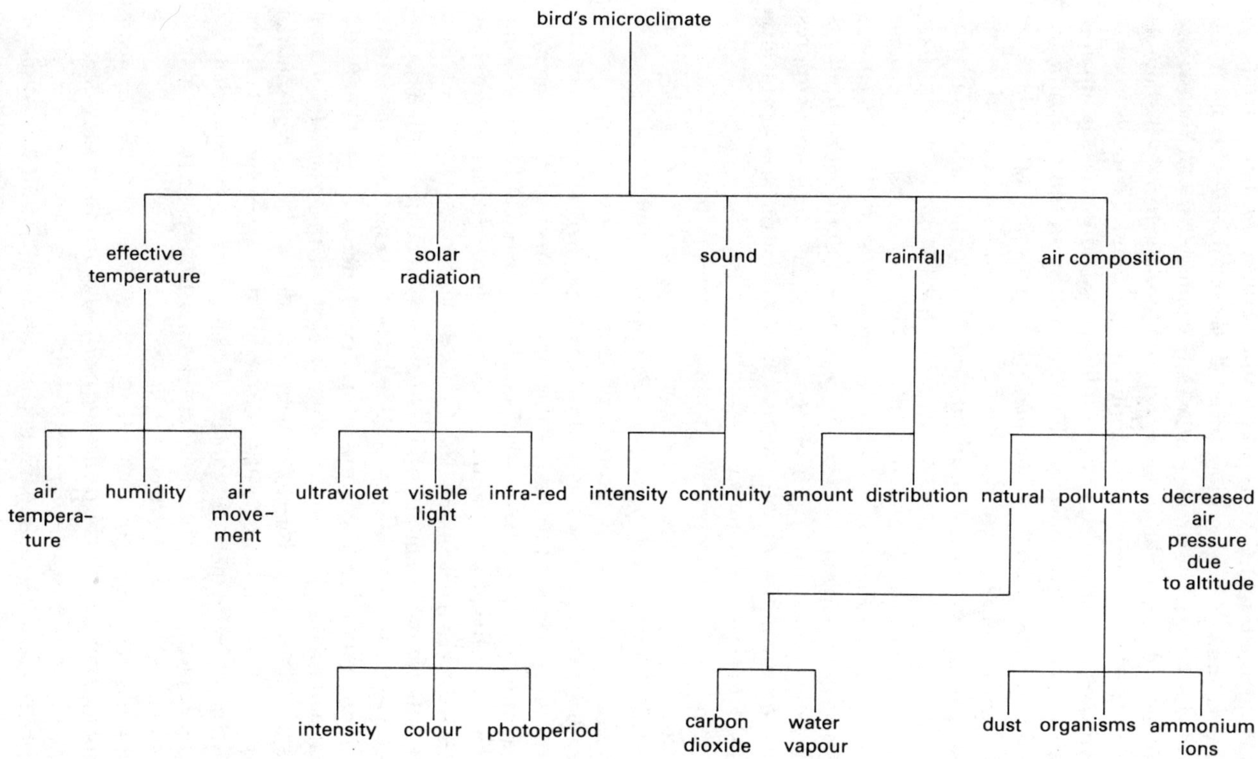

Fig 3.1 *Factors affecting the microclimate in which the birds live*

Stresses caused by high temperature and humidity

The internal body temperature of domestic poultry is normally 41.2–42.2°C, considerably higher than that of mammals (36–39°C). Newly hatched birds have a body temperature approximately 2–3°C below that of the adult bird and the adult temperature is not reached until about six days after hatching. The reason for this is that, as chicks develop, their body surface area per unit body weight decreases and this is accompanied by a progressive improvement in the efficiency of thermal insulation of the body as fat is deposited and feathers develop. Consequently there is a gradual increase in body temperature, and reductions in the lethal ambient temperature, critical temperature and the optimum temperature for growth.

The fundamental problem facing domestic poultry exposed to extremes of temperature is to maintain their body in a condition which permits the normal functioning of the chemical processes. Death from cold occurs well above the point at which freezing of the tissues begins, presumably because the rate of energy production falls below the minimum level necessary for basic maintenance.

The upper temperature at which life can exist is that at which the denaturing of protein (including enzymes) occurs. However before this happens death often takes place because of desiccation.

The optimum body temperature in birds is much closer to the point of heat death than of cold death, and domestic poultry maintain a body temperature which is only 4–5°C below the point at which enzyme inactivation begins. In contrast, the body temperature of birds can drop by 20°C below normal body temperature, and then if rewarmed carefully the birds will recover completely.

Fig 3.2 shows the idealised relation between metabolic rate and temperature. The range of temperature in which metabolic rate is at a minimum is known as the **zone of thermal neutrality**. The actual temperatures at the lower and upper extremes of this zone depend on insulation of the bird and its level of feeding. Below the zone of thermal neutrality food is used wastefully and above this zone the bird suffers heat stress. Birds are more cold tolerant than heat tolerant and they are much more likely to die from heat stress than from cold stress.

The interaction between ambient temperature and nutrition

With a few exceptions it is normal practice to offer food *ad libitum*[1] to laying birds. Under these circumstances the energy content of the diet is

1 To the extent desired

predominant in governing intake. The energy intake mechanisms are certainly sensitive to environmental temperature because daily food intake by laying hens has been shown to decline as environmental temperature is increased. The decline in intake is curvilinear resulting in a reduction of approximately 1.5% in appetite per degree between 21°C and 30°C and 4.6% per degree between 32°C and 38°C.

It is clear from Fig 3.2 that heat production by the bird is lowest in the zone of thermal neutrality but increases outside this zone particularly in cold ambient temperatures. At high environmental temperatures just below the bird's body temperature, birds are unable to lose all the heat they are producing by evaporation. Consequently body temperature rises and as a result heat production also increases. This process rapidly leads to their death from hyperthermia.

Increased heat production has implications as far as efficiency of food

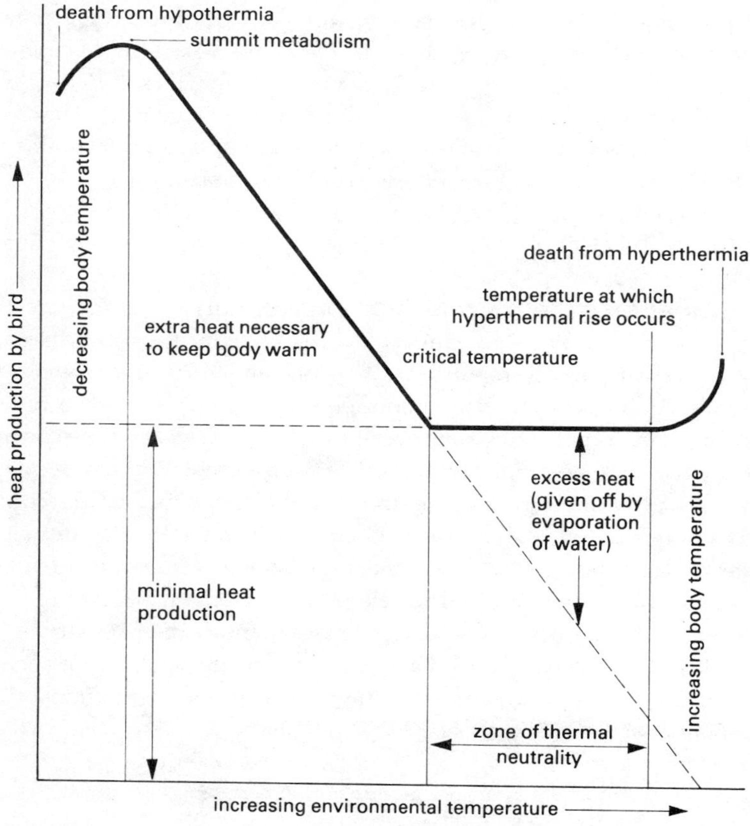

Fig 3.2 *Effects of environmental temperature on body temperature in poultry*

utilisation is concerned. Fig 3.2 refers to fasting heat production but heat is also produced as the result of exercise, production and the ingestion of food. Heat produced by these means is an advantage to the bird in cold conditions but a liability in hot conditions since it adds to the heat stress imposed on the bird. Consequently it is not surprising that birds react to high temperatures by reducing their food intake.

In fact poultry are not particularly well adapted to high ambient temperatures since their skin contains no sweat glands and only the head appendages are very vascular. Therefore little evaporative loss occurs from the body although there is some diffusion of water through the skin. In warm conditions chickens allow water to drip from their facial appendages to encourage evaporative cooling, and these appendages and the legs are warmed by vasodilation to encourage sensible[1] heat loss. However most of the evaporative heat loss in birds occurs from the lungs. The lungs of poultry are small relative to body size compared with those of mammals but they have at their base large non-vascular air sacs which are not involved in gaseous exchange, but allow extensive areas of moist surface for evaporative cooling. The birds lose heat by panting. Panting occurs when the deep body temperature reaches 43°C but even before this respiration deepens and evaporation increases. Above the panting threshold respiration frequency increases but tidal volume decreases. The respiration frequency reaches a maximum of 140–170 respirations per minute.

Effect of ambient temperature on productivity

Most highly productive poultry are kept in temperate zones where the effect of cold stress is likely to be more important than the effect of high ambient temperatures. However over the last twenty years poultry scientists have become more interested in the effects of high ambient temperatures on poultry and especially on modern hybrid layers for two reasons.

- It has become clear that the food efficiency of poultry (in particular layers) can be improved by keeping them at temperatures higher than those normally found in temperate regions in winter.
- More and more tropical countries have expanded their poultry industries and as a consequence have encountered the problem of keeping modern hybrid poultry both alive and productive under these conditions.

1 Sensible heat loss (conduction, convection and radiation) warms up the surrounding environment whereas insensible heat loss (evaporation) does not.

Much of the work that has been carried out has been undertaken in climate rooms. This work although valuable has led to conclusions which rather oversimplify the problem.

Effect of ambient temperature on the productivity of laying birds

Much of the work on laying birds has been undertaken on white breeds of birds and very little on brown breeds. Therefore few objective results are available on the relative merits of these two different types with regard to their tolerance to high ambient temperatures.

Effects of ambient temperatures on energy intake and production of eggs

It has already been reported that food intake by laying birds declines exponentially as environmental temperature is increased. Consequently it is not surprising that many research workers have found that at high ambient temperatures a reduction occurs in the number of eggs produced by laying hens.

The decline in productivity by laying hens at high ambient temperatures is probably caused by a combination of heat stress *per se* and an insufficiency of certain nutrients in the daily intake. However, experiments have shown that the main cause of loss of production was in fact the reduction in energy intake caused by the high ambient temperature. Thus the adjustment of diets to increase their levels of energy will have little effect on productivity. Although it is desirable to ensure that the intake of essential minerals, vitamins and protein is adequate at these low levels of intake, the consumption of these by hens in the absence of adequate energy intake will not be sufficient to ensure that they maintain their rate of egg production (Table 3.1).

Diets should be formulated so that laying hens consume 15 g of high quality protein containing at least 0.5 g of sulphur-containing amino acids each day. When such diets are fed, egg production should not be affected at ambient temperatures as high as 32°C. The effect of high ambient temperatures on egg production appears to be progressive and consequently hens kept at ambient temperatures above 32°C show a progressive decline in egg production.

Table 3.1 The effect of environmental temperature on energy intake and heat production by White Leghorn laying pullets

Temp (°C)	Estimated daily energy intake/kg metabolic body weight (kJ)	Estimated daily heat production/ kg metabolic body weight (kJ)	Difference i.e. energy available for egg production/kg metabolic body weight (kJ)	Energy available for egg production converted to 1.489 kg metabolic body weight (1.7 kg body weight) (kJ)	Possible production of 57 g eggs by 1.7 kg pullet (assuming 380 kJ per egg) (%)
26.5	819	610	209	312	82
29.5	798	597	201	299	79
32.0	727	551	176	263	70
35.0	614	480	134	202	53
38.0	452	374	78	115	31

Smith (1970)

Effect of ambient temperature on egg weight

It is well known that laying birds kept at high ambient temperatures produce eggs of lighter weights than similar birds kept at lower ambient temperatures. The effect of constant ambient temperature on egg weight seems to occur at a lower ambient temperature than the temperature which reduces egg production (32°C). The reduction in egg weight (80%) is caused mainly by the direct effects of ambient temperature on the physiology of the bird and only slightly (20%) indirectly through food intake. Thus although it is possible to increase average egg weight by feeding higher energy diets, especially those containing fat, it is not possible to offset completely the effects of high ambient temperature. In fact high energy diets increase the mean weight of eggs produced from hens kept at moderate ambient temperatures (15–26°C) more than they increase the mean weight of eggs produced by birds kept at higher temperatures (27–35°C). The effect of ambient temperature on average egg weight appears to be cumulative (Fig 3.3).

Thus when birds were kept at 26°C their mean egg weight increased by 1 g per week whereas when kept at 35°C the average egg weight remained constant for a six month period.

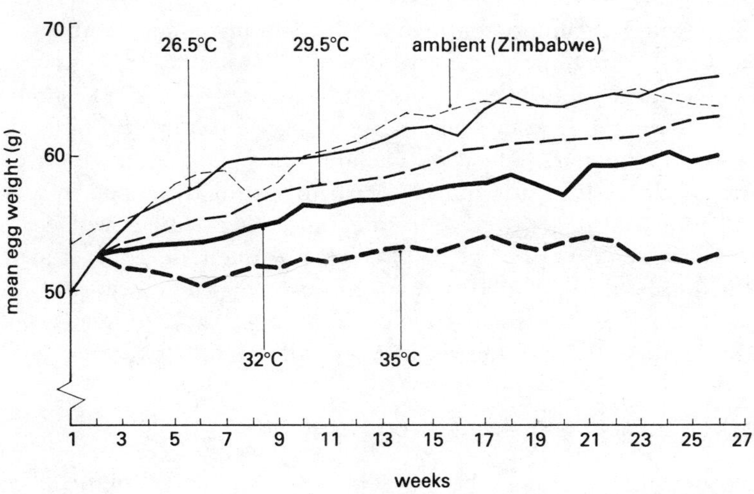

Fig 3.3 *Changes with time in the mean weight of eggs produced by pullets which were kept at different environmental temperatures after an initial two weeks at 26.5°C (weeks 1 and 2 on graph)* (A.J. Smith 1970)

Effect of ambient temperature on shell weight

The effect of temperature on the weight and thickness of shells produced by laying hens is immediate. The first egg formed after exposure to temperatures above 35°C is thinner than normal. Eggs produced by birds kept at 38°C have such thin shells that 100% of them break. The decline in shell thickness, which occurs above the panting threshold, is almost certainly due to the reduced ability of blood to carry calcium ions, a phenomenon associated with alkalosis resulting from panting. The recovery of shell thickness after the removal of heat stress is as rapid as the decline. The first egg formed after the removal of the heat stress is likely to be almost normal.

Neither feeding extra calcium nor feeding extra vitamin D or vitamin C seem to have much effect on the thickness of shells. A trial carried out in Côte d'Ivoire in which ascorbic acid (vitamin C) was added to the diet of laying hens confirmed this latter conclusion. Since the effect of ambient temperature is most severe when the ambient temperature is raised above the panting threshold, the best way to cure this problem would seem to be to modify the environment to avoid the birds becoming stressed.

One solution to the overheating that occurs during the hottest part of the day (the daylight hours) is to reduce heat production during this period. Research has shown that heat production during the day is about 35% higher than it is at night. This means that the bird is producing most heat at the very time when it has most difficulty in losing it and when it will be gaining considerable heat from its surroundings. The normal feeding practice of the bird makes the situation worse since most food intake occurs during the hours of daylight. However this need not necessarily be the case. At the Poultry Research Centre in Edinburgh it has been found that if food is available **only during the last three hours of daylight**, birds quickly learn to eat all their food in a three hour period and they eat the same amount of food as they would if food were available during the whole day. This regime reduced daytime heat production by more than 16%, and increased nightime heat production by 18%. Food intake would have to decrease by 50% to give a similar fall in daytime heat production in hens with continuous access to food.

Effect of ambient temperatures on water intake

The water intake of laying hens is substantially modified by the effect of high ambient temperature. When laying birds are exposed to a sudden increase in ambient temperature above the panting threshold they more than double their water intake (from 250 to 500 g a day). However this increase may not be maintained for more than twenty-four hours. Most birds will then reduce their water intake to the levels they were consuming prior to the increase in ambient temperature. This decrease in water intake is associated with a decrease in food intake and a reduction in the rate of egg production. However the ratio of water intake to food consumption remains substantially higher than at temperatures below the panting threshold.

Not all birds behave in the way outlined above. Some individuals maintain water intake, food consumption, and egg production at high levels and it is possible that these birds could be used as the basis for the selection of a strain for use at high ambient temperatures. A few birds behave in a peculiar way, drinking large quantities of water (1000 g a day) which appear to pass straight through the birds' intestines. This mechanism cannot be efficient for cooling birds since the only really effective heat loss mechanism is the evaporation of water from the lungs.

Effect of fluctuating ambient temperatures

The effect of fluctuating temperatures on the productivity of laying hens is largely dependent on the mean temperature. When constant temperatures are compared with fluctuating ones, if the mean is below the panting threshold, say 29°C, fluctuating temperatures are likely to have the same effect as constant ones in terms of food intake, rate of lay and egg weight.

If the average ambient temperature is above the panting threshold the birds kept on the fluctuating temperature are likely to have an advantage over those kept at a constant temperature.

The importance of a recovery period cannot be over-emphasised and laying birds can withstand very extreme ambient temperatures (up to 47°C) for two to three hours provided that the temperature then declines and they are kept at an ambient temperature of not more than 21–26°C for several hours to allow them to lose the heat stored over this period. Laying birds kept under these conditions have even laid eggs at more or less normal rates.

Breed and bird differences

Both brown and white strains of laying hens can be productive at high ambient temperatures, although white birds tend to be slightly more productive. What is interesting is that the effects of high ambient temperatures affect individual birds very differently (Table 3.2), and consequently at high ambient temperatures the coefficient of variation for egg production is over 100% (compared with 12–14% in birds kept at 21°C).

Table 3.2 **Variability in marketable egg mass production between individuals within two breeds of pullet held in the different temperatures and recorded over 20 weeks**

Temperature		**Number of pullets that produced marketable eggs at the following rates (gram of egg/bird/day)**				
		>45	**45–35**	**34–25**	**24–15**	**<15**
18°C	Ross Apollo	10	4	1	0	0
	Ross Ranger	8	8	0	0	0
32°C	Ross Apollo	0	6	6	0	0
	Ross Ranger	0	3	3	4	5
32°C	Ross Apollo	0	3	3	6	2
70% RH	Ross Ranger	0	0	0	5	9
35°C	Ross Apollo	0	1	1	1	8
	Ross Rangger	0	0	0	5	8

After Kumar (1977)

Although all birds are affected to some degree by exposure to high ambient temperatures, some birds are affected more than others. This may mean that there is sufficient genetic variation within populations of modern hybrids to allow for selection for productivity at high ambient temperatures. A start should be made to identify the physiological differences between the high producing and low producing birds.

Effect of ambient temperature on the productivity of growing birds

Effect of ambient temperature on growth and food conversion

The growth rate and food intake of broilers aged between three and eight weeks is affected in a linear manner by ambient temperatures between 7° and 21°C. Both growth and food intake decreases by 0.12% for each degree rise in environmental temperature over this range. Above this range of temperature both the growth rate and food intake are affected in a curvilinear manner resulting in a rapid increase in the time birds take to reach slaughter weight. Food conversion efficiency is at a maximum in the range of temperature 21–26°C.

Effect on carcass composition

At higher ambient temperatures broilers grow more slowly than at lower ones; therefore it follows that at a given age the total meat production (g/bird) is reduced. Total carcass fat and abdominal fat increases by about 0.8% and 1.6% respectively with each degree rise in temperature. Carcass moisture declines with increasing ambient temperature but there is no significant effect on carcass protein (expressed as a percentage of live weight). However there is evidence to suggest that in males there is a linear reduction in the amount of breast meat produced by males as the ambient temperature at which they are kept is increased.

Interaction between temperature and nutrition

At high ambient temperatures the less energy a broiler has to lose the better. Thus any method of feeding that lowers heat increment should be advantageous. Therefore the feeding of broilers on high fat diets should be an advantage because fat gives a lower level heat increment than other energy foods. However results achieved when this theory has been put to the test have not been conclusive.

Similarly because of the reduced intake of broilers at high ambient temperatures the concentration of amino acids in the diet should be increased. The relative balance of each amino acid should be carefully considered because imbalances can lead to extra heat production which could be a problem with broilers kept at high ambient temperatures.

Effect of humidity of the air

The effect of the humidity of the air is very much interrelated with the ambient temperature. At 21°C, relative humidity within the range 48–90% has no effect on growth or food conversion but at 29°C increasing the humidity from 30% to 70% has been shown to slow the growth rate of broilers.

Modifying the environment

Introduction

When poultry are kept in the tropics they are often exposed to thermal stress which can lead to a considerable loss in productivity as has been demonstrated earlier. This loss in productivity is particularly important with high producing birds. High producing birds are normally housed in large purpose-built houses but unfortunately the design of these houses is often copied from designs developed for use in temperate countries which are largely unsuitable for tropical countries. In general they are expensive to build and run and do not provide sufficient insulation or airflow. The important aim is to achieve a microclimate which will enable poultry to maximise their productivity.

The house must provide the birds with an equable temperature with limited diurnal fluctuations. It must keep the birds dry, be easy to manage, easy to clean, and with laying birds, have the equipment necessary to allow the day-length to be controlled. These conditions have to be provided under two broad climatic zones, hot humid areas and hot dry areas.

Hot humid areas lie roughly 15°S–15°N. The air temperature in these regions ranges from about 24–35°C with a mean temperature of 26°C. The humidity of the air is high and the air movement usually very low.

> *High humidity in the chicken house is an important cause of deaths of young poultry during the rainy season.*

Hot arid areas lie roughly 15–30°N, and 15–30°S. They experience a much wider range of temperature than the hot humid zones, 7–48°C. This fluctuation occurs diurnally and seasonally. For example in Khartoum the temperature range is from 7–40°C in January and 20–48°C in June. In this type of climate there is also a greater intensity of solar radiation and more air movement than in the hot humid zone.

Siting of poultry houses

Houses should be sited to take advantage of topographical features which will favour air movements. Records of local wind speed and direction should be studied to ensure the best orientation to take advantage of prevailing breezes. Siting of buildings close to plantations or other buildings should be avoided so that air movement through the building is not obstructed.

Tall trees near to the buildings are an advantage (Fig 3.4) because these provide shade without obstructing airflow, whereas shrubs will provide little shade but will obstruct airflow.

The proximity of trees and vegetation can contribute to cooling as a result of leaf transpiration. Wind blowing across crops in the dry tropics can be cooled by several degrees Celsius.

Fig 3.4 *Poultry house shaded by trees*

It is sensible to site poultry houses close to irrigated crops especially grassland. Also if the house is surrounded by grass or another cover crop less solar radiation will be reflected into the house than it would be by bare ground.

In the case of housing for ducks an area of shallow water outside the house can enable the ducks to tolerate air temperatures that would be fatal to other poultry. This area of water needs to be only half the area of the house and only about 15 cm deep.

The building

The roof is the most important part of a house in a tropical country. It should be of such a design that it effectively intercepts solar radiation and prevents the birds being exposed to direct solar radiation (Fig 3.5). Simply giving shade can reduce heat loading on the hen by 30–45%.

Ensuring that the birds are always shaded is a first priority in a hot climate.

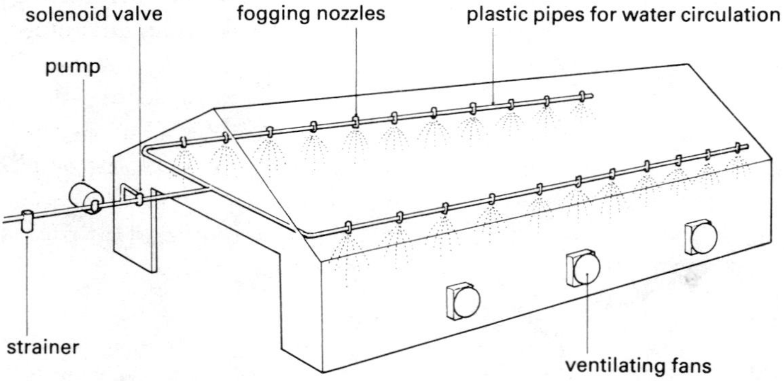

Fig 3.5 *High pressure misting (fogging) evaporation cooling system in a poultry house* (courtesy of NRG Company, P.O. Box 306, Ardmore, PA 19003)

Heat stress on the birds is lessened if they are surrounded by cool surfaces. Insulating the roof is beneficial as this reduces the long-wave radiation down to the birds. The type of insulation depends on cost and what is available. It must however be borne in mind that although thatch is a good insulator it can also harbour vermin and parasites. For these reasons it is probably advisable not to use this material. Non-insulated corrugated iron roofs should be avoided since they transmit a considerable proportion of the incoming solar radiation (Fig 3.6).

Fig 3.6 *An undesirable type of poultry house constructed from corrugated iron sheets*

The amount of solar radiation absorbed by a surface depends on its colour and structure. White or shiny smooth surfaces are cooler than similar dark rough surfaces. Thus white paint on the upper surface of the roof helps to keep it cool. One of the coolest areas of the surroundings is the sky, thus maximum exposure of poultry to the sky is an advantage. It has been shown that the temperature of a cloud-free sky can be about 16°C cooler than the air when the temperature of the air is 38°C and about 12°C cooler when the air temperature is 43°C. Thus in tropical latitudes south of the equator it is advantageous to have the southern section of the roof located a considerable height above the birds so as to ensure a good exposure to the southern sky, providing this does not make the building too susceptible to damage by wind.

In most cases houses are best designed with an open-sided structure to promote air movement. Natural ventilation is cheap and generally reliable.

Air movement removes sensible heat, moisture and pollutants. For birds in battery cages, a flow of air through the house is essential to prevent the birds in the upper cages becoming too hot. Some form of wall such as wire netting may be needed to keep the birds in and to keep predators and thieves out. In some climates, one solid wall may be necessary to protect the birds from strong winds and rain; in other cases temporary protection should be provided at certain times of the year.

Buildings with a large thermal mass absorb heat during the day and release it at night thereby reducing temperature variation within the

Fig 3.7 *Enclosed insulated house in Saudi Arabia*

structure. This property is relatively unimportant in poultry houses which have free access to the outside air since the temperature inside the building follows the temperature outside the building. It is only when there is a large diurnal variation in temperature and the airflow through the building is at a very low level that high heat capacity and low thermal conductivity walls can be fully and satisfactorily applied. Therefore this type of building can be used for small numbers of birds in the dry tropics. It is however rarely used for large poultry houses in the tropics as a high degree of ventilation is needed to prevent a build-up of ammonia, particularly in the humid tropics. If atmospheric ammonia exceeds 500 ppm both respiration rate and food intake are reduced resulting in decreased egg production.

In the tropics poultry houses are often too sophisticated and consequently too expensive. Sophisticated buildings can be justified only if the price structure is correct, as for example was the case in Saudi Arabia in the early 1980's (Fig 3.7).

Both the initial costs and the running costs of controlled environment buildings are very high, and they depend on a reliable supply of electricity and maintenance services. If the cooling and ventilation system breaks down, the temperature inside a densely-stocked enclosed poultry house rises rapidly with fatal consequences. Therefore an alarm system, a stand-by generator, or a fail safe device which opens up the house to allow natural ventilation are essential.

4 Poultry nutrition

Digestion

The digestive system of the fowl is simple but very efficient (Fig 4.1). Food is picked up by the beak and selected on the basis of feel and appearance rather than taste. There is little or no evidence that chickens or other poultry can smell or taste.

The food, which is swallowed whole, passes quickly through the oesophagus to the crop which acts mainly as a storage organ (Fig 4.1). In this organ the food is softened and acidified by lactic acid produced by bacterial fermentation. The food passes from the crop to the proventriculus by means of a short connecting tube where the food is acted upon by enzymes, particularly pepsin, and further acidified by hydrochloric acid. The proventriculus is the organ which most closely resembles the mammalian stomach. The gizzard is just beyond the proventriculus; it is a strong muscular organ which crushes the food by rhythmic contractions into a pulp. This process is assisted by the presence of insoluble grit. This hard grit is particularly valuable when whole grains are fed to birds but is not necessary when mash feeds are used. The food then passes into the duodenal loop which is the first part of the small intestine. This loop surrounds the pancreas which secretes pancreatic juices into the duodenum. Slightly lower down the intestine, the liver is joined to the small intestine by two ducts one of which comes from the gall bladder, which stores bile salts. The lining of the duodenum and the small intestine is convoluted into finger-like projections (the villi). These increase the surface area of the gut and thereby aid absorption of food. The food is moved through the small intestine by regular peristaltic contractions. It is in this region of the gut that most of the digestion and absorption of food takes place. At the junction of the small and large intestine are two blind sacs known as caecae. Their main function is to digest fibre and absorb water. The large intestine is also responsible for the absorption of water. From the large intestine the faeces pass to the cloaca where they are evacuated.

The digestive system of the duck is similar to that of the fowl. The bill of the duck is much larger than that of the fowl but the crop is smaller. When feeding from a trough ducks tend to push food out of it and onto

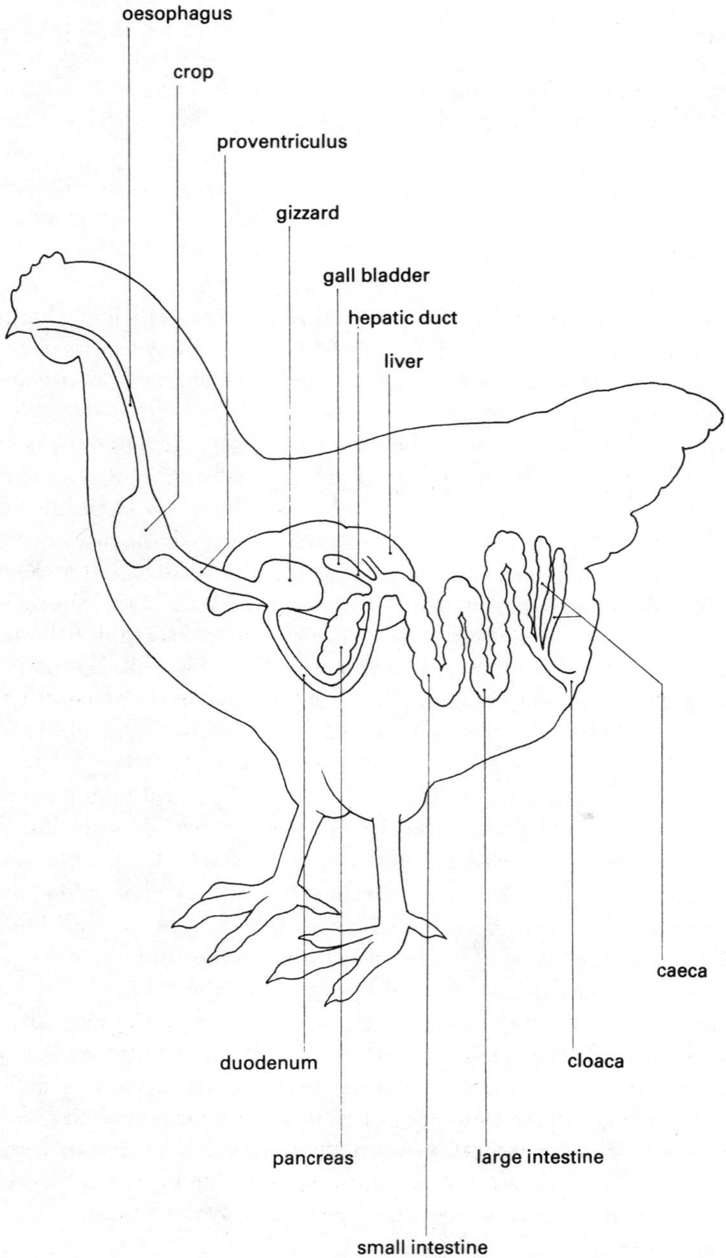

Fig 4.1 *Digestive system of the fowl*

the ground causing considerable wastage. Unlike the fowl, ducks have a small intestine that can absorb salt when the dietary intake of this mineral is high. This excess salt is then excreted by nasal glands. Fibre can be fermented in the caecae of ducks to produce volatile fatty acids. It is believed that improved digestion of fibre can be achieved by encouraging these organs to increase in size by feeding ducks diets containing high amounts of fibre.

The goose has no crop for grinding up food but there is an enlargement at the end of the gullet that serves as a temporary food storage organ.

Nutrient requirements

Most of the research work on domestic poultry has been carried out on domestic chickens (*Gallus domesticus*). Thus most of the information on nutrition has been gathered for this species. Where differences between *Gallus domesticus* and the nutrition of other domestic poultry exist those differences will be pointed out. Otherwise the section on general nutrition can be taken to apply to all domestic poultry. Because poultry are monogastrics they are unable to manufacture essential amino acids or the B vitamins, and they cannot exist on high fibre diets. The diet of birds which are intensively housed and which have access to neither soil, grass, nor sunshine must contain the materials essential for the processes of maintenance, production, and reproduction. These essential nutrients can be conveniently grouped under the following headings:

- water
- carbohydrate } sources of energy
- fats and oils }
- protein (amino acids)
- vitamins
- minerals

The following sections consider these nutrients.

Water

Water is normally provided *ad libitum* for poultry and is therefore often not given its due consideration by nutritionists. Water intake increases with age (Table 4.1).

> *A lack of water can seriously retard growth and impair egg production. This is particularly true in tropical countries where deprivation can lead to death of poultry within a very short period of time.*

The body of a bird contains in the region of 60% water, and eggs have approximately 65%. The presence of water is essential if nutrients are to be absorbed and toxic materials removed from the body.

Table 4.1 Estimate of water intake by chickens of various ages in a temperate environment

Age (weeks)	Water intake per 100 birds (litres/day)
0–2	4–5
2–5	7–10
5–10	15
10–20	18
Adult layers	20–30

Water is also essential for the control of body temperature by the bird in hot environments. Panting is an essential heat loss mechanism under these conditions and lack of water quickly leads to death by hyperthermia. Birds consume considerably more water at high ambient temperatures than at low ambient temperatures. The other factors which influence water intake include diet, rate of lay and size of bird.

Effect of temperature on water consumption

The water intake of laying birds will initially increase rapidly when ambient temperatures rise above 21°C. However this high level of intake will be maintained for only the first two days of heat stress. With continued exposure to high temperatures water intake by hens returns to its normal level. This reduction of water intake is accompanied by a fall in food consumption by birds.

Effect of diet on water requirements

The ratio of dry matter intake to water intake by poultry in a temperate environment is normally 1:2 but does depend to some extent on the diet. Increased protein level increases water consumption and the ratio widens but the effect is not great. Sodium chloride in the diet increases water intake, and in one experiment water intake per gram of food consumed increased from 1.70 cm^3 on a diet containing 0.18% salt to 8.71 cm^3 when 8% salt was added to the diet. Potassium and magnesium salts can have a similar effect on increasing water intake.

Effect of water restriction

It is undesirable to restrict water intake of any bird, particularly in the tropics, where deprivation can lead to death within twenty-four hours. Even a 10% restriction of the amount of water available can reduce the growth rate and food conversion efficiency of broilers. With layers the effect is even more devastating. Short periods of deprivation can result in the birds moulting and ceasing to produce eggs.

Energy requirements

The energy in foodstuffs fed to poultry is normally expressed in units of metabolizable energy per unit weight, e.g. kilojoules per gram (kJ/g) and the requirement of poultry in terms of metabolizable energy per day (kJ/d). The **metabolizable energy** refers to that portion of the food which is available to the bird for the production of meat and eggs and for the maintenance of vital functions and of body temperature (Fig 4.2).

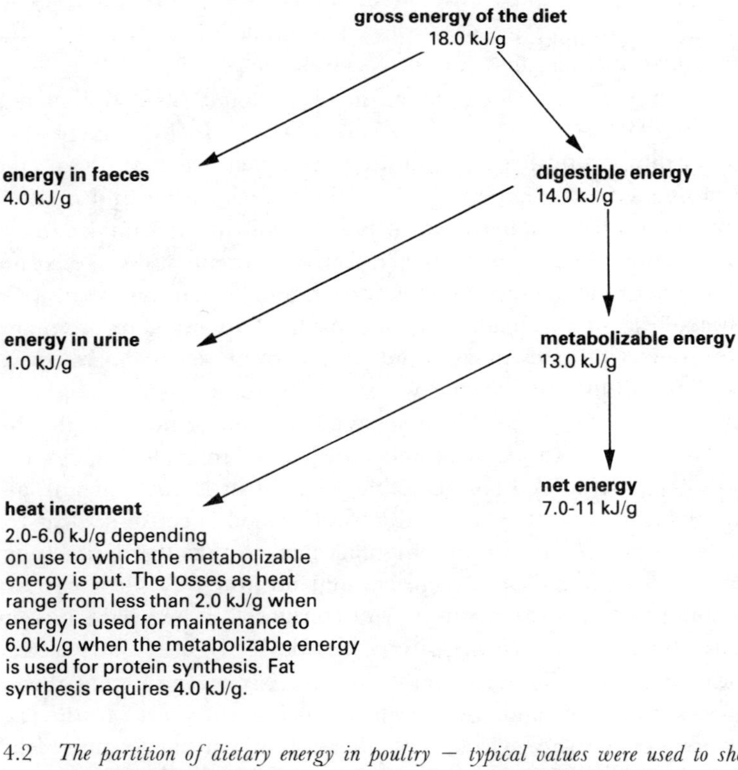

Fig 4.2 *The partition of dietary energy in poultry – typical values were used to show quantitatively the loss of energy from a diet via the different routes in chickens*

The energy in the diet is largely in the form of carbohydrates but some is fat and some amino acids. In most cases poultry are given free access to food and the birds are allowed to consume as much food as they wish. Poultry usually consume just enough food to meet their nutrient requirements. This control of intake is based primarily on the amount of energy in the diet.

Birds eat to satisfy their requirement of energy.

Thus increasing the concentration of energy in the diet decreases intake, and vice versa. This is true provided that the diet is adequate with regard to all other essential nutrients, and that bulkiness, texture, inaccessibility and palatability do not limit intake. Therefore when a diet is being formulated the levels of the constituents are stated in relation to the energy content of that diet. It is not normal to distinguish between maintenance and production requirements of poultry as is the case with mammals.

Voluntary food intake

Appetite or voluntary food intake is the amount of food that a bird consumes when it has unlimited access to a diet. Knowledge of a bird's appetite is important because if a flock consumes less than the predicted level of intake it may experience nutritional deficiencies and production will be below that predicted. Also a reduction in food intake can be a useful warning of errors in management or outbreak of disease. If intake is above the predicted level food is wasted and certain types of bird may become obese. The mechanisms controlling appetite are complex – overall control of food intake is influenced by the hypothalamus in the base of the brain, but the pathways that transmit instructions are still incompletely understood. Current hypotheses assume that the hypothalamus responds to physical and chemical changes in the circulating fluids of the body. The body temperature of birds is not constant, and it has been suggested that heat produced after food is consumed raises the temperature of the blood and hypothalamus, so that the desire to eat is lessened. This thermostatic theory would explain why birds eat less at high ambient temperatures and why food with a low heat increment such as fats can cause obesity.

An alternative is the glucostatic theory. This proposes that there are glucose receptors in the hypothalamus which are sensitive to the rate at which glucose is being utilised by them. Low utilisation rates lead to hunger sensation.

The above theories can possibly explain short-term regulation of food intake; long-term regulation is probably concerned with the prevention of excess fat deposition. According to this theory the hypothalamus is sensitive to concentrations of circulating metabolites mobilised from endogenous fat stores. Since the amount of fat mobilised is proportional to the size of fat deposits, a lipostatic mechanism keeping the fat content constant, could control body weight.

These then are the main hypotheses regarding the physiology of food intake in birds. No one hypothesis alone describes adequately the observed food intake behaviour. Therefore it is likely that multiple factors are responsible.

Factors which influence voluntary food intake

The most important factors affecting food intake are the characteristics of the bird (body weight, rate of liveweight gain, output of eggs), quality of food, and the environment.

Body weight

Heavy birds consume more food than light birds. The control of food intake of light-bodied laying hens is very precise and they adjust to considerable variation in the energy levels in the diets and will maintain their daily intake of energy at a constant level. Heavy birds, on the other hand, will consume more energy on a high energy diet than a low energy one and become obese.

Rate of gain

Birds that grow faster than the average normally consume more food than the average. Faster growth means better food conversion, because a greater proportion of the food is used for production. At a given weight, immature birds of broiler strains consume more food than birds of an egg laying strain.

Output of eggs

The intake of dietary energy by laying hens is related to their rate of egg production. A 1% increase in egg production is associated with a 2% increase in food intake. Food intake is greater when active egg formation is taking place and there are large differences in intake by hens between egg forming days and other days. Laying birds have been shown to consume 20% more food on egg forming days than days when eggs are not formed. Food intake increases markedly when birds go through the hormonal change two weeks prior to the onset of egg production.

Dietary effects

The major dietary factor which affects food intake is the concentration of energy in the diet. An increase in dietary energy results in a decrease in food intake. The concentration of nutrients other than energy in the diet does not influence appetite so long as the level does not go outside the range which is acceptable for normal health and production. If the diet is deficient in one or more essential nutrients, appetite is depressed and this is associated with a decline in growth or reproductive performance. In the case of a marginal deficiency of an amino acid, for example methionine, both hens and chicks will sometimes consume extra food to restore their intake of the limiting amino acid to an adequate level. In this case an appetite for the amino acid overrides the appetite for energy. The effect is a small one and occurs only when the concentration of the amino acid in the diet is just below the level needed for maximum performance.

The intake of pelleted food can be up to 8% greater than the intake of the same food presented as a meal. This is partly an effect of partial cooking of the food and partly because the pellets are more convenient to consume. This high intake of pellets is desirable in young birds since it promotes growth rate, but in the adult it can lead to increased fat deposition and obesity.

Other factors affecting intake

Light intensity and day-length have some effect on voluntary food intake. With laying birds this is an indirect effect of day-length on egg production. Longer days stimulate egg production and therefore encourage hens to consume more food. With chicks, light has a more direct effect upon food intake by regulating behaviour patterns and extending food activity. Maximum food intake and growth rate are obtained when chicks are reared in continuous light. With broilers too much light may increase activity and therefore reduce efficiency of food utilisation.

Ambient temperature also has a major effect on food intake but this is dealt with in Chapter 3.

Restriction of energy intake

Pullets of egg producing strains

It is not the usual practice to restrict the food intake of growing pullets being reared for egg production. However, there is considerable evidence that food restriction is beneficial. The main advantage is a saving in the amount of energy required to reach point of lay. In addition, restricted feeding reduces body weight at point of lay, delays the onset of sexual maturity and decreases laying house mortality, but it does increase mortality during rearing.

A common level of restriction has been 70% of *ad libitum* for the period eight to twenty-one weeks of age.

Some recent evidence indicates that restriction of food intake during the laying period by 5–10% might improve efficiency of food conversion. However, it is not normal practice and it is a rather risky procedure because a comparatively small restriction can cause a dramatic drop in egg production.

> *Although restricted feeding delays sexual maturity, subsequent egg production is normally increased and restricted birds generally show a higher peak of rate of lay with a slower rate of decline thereafter.*

Pullets of meat producing strains

Pullets of heavier stock are not normally fed *ad libitum* during the growing stage but are restricted in their food intake so as to reduce liveweight gain and in particular to limit the amount of body fat at the onset of egg production. The amount of restriction depends upon the particular genotype involved but the level of restriction is much more severe than that applied to birds of laying strains.

Various methods of restriction of food intake can be used. These include skip-a-day feeding, low protein diets, high fibre diets and low lysine diets. Skip-a-day feeding is probably the system of choice. With this system if birds are to be rationed to 70% of their voluntary intake, they are fed 140% of their daily intake every two days. This technique ensures that every bird, including those low down the pecking order, is able to obtain its fair share of food. If birds are rationed to 70% of their food intake every day birds high up in the pecking order will achieve 100% of their desired consumption, whereas birds lower down the pecking order may consume less than 50%. With birds of modern broiler strains much more severe levels of restriction (less than 50% of appetite) are applied, particularly in the rearing phase.

> *The best advice for any farmer keeping hens of these strains is to follow the advice of the supplier of the chicks.*

One other point, if birds are given food in a restricted way and water is provided *ad libitum*, wet litter may result in the humid rainy season. This problem can be overcome by restricting the water intake of the bird.

Protein requirements

Foods containing high levels of protein are expensive to purchase, and so a diet which is too high in protein is unnecessarily expensive. The excess protein is broken down and used as an energy source and the excess nitrogen is excreted as uric acid. The synthesis of protein in the body tissues requires an adequate supply of about twenty different amino acids. Ten of these cannot be synthesised by the bird and must therefore be provided in the diet (Table 4.3). These ten amino acids are termed **essential**. Of the remaining ten, three can be readily synthesised by the body, these are termed non-essential. The remaining seven are less easily classified because either they can be formed only from essential amino acids or the rate at which they are formed is not always sufficient to meet the demand of the productive processes.

Table 4.3 Classification of amino acids

Essential Lysine, methionine, threonine, tryptophan, isoleucine, leucine, valine, phenylalanine, histidine, arginine.
Non-essential Alanine, aspartic acid, glutamine.
Difficult to classify Tyrosine*, cysteine**, glycine***, serine***, proline.****
* Tyrosine – can be formed from phenylalanine. ** Cysteine – can be formed from methionine. *** Glycine and serine – are interconvertible, but synthesis is sometimes inadequate. **** Proline – rate of synthesis is sometimes inadequate for body functions.

There are no indications of differences in the digestibility of protein by chickens as compared with ducks or other domestic poultry.

Definition of a protein requirement

The protein requirement of a bird is defined as a requirement for a supply of each essential amino acid together with a sufficient supply of suitable nitrogenous compounds from which non-essential amino acids can be synthesised. It is unsatisfactory to specify a protein requirement as a single figure, since the amount of protein which must be supplied depends upon the yields of amino acids obtained by the bird when that protein is digested.

Two different protein sources may have low biological values when fed separately, but when fed together they may act synergistically and have a much higher biological value. **Biological value** is defined as the single numerical value ascribed to a protein which gives an estimate of the relationships between the essential amino acid balance of the diet protein and the amino acid needs of the bird. If the biological value is eighty, then for each 100 g of protein absorbed 20 g are judged not to be useful because of amino acid imbalances. An example of such a synergistic relationship exists between the protein from groundnut meal and that from sunflower meal. The reason for the higher biological value is that the amino acid profile in one of the proteins complements, at least in part, the amino acid profile of the other. A second point is that the amino acid requirement of different metabolic processes such as growth and egg production are not the same. Therefore an amino acid profile which will meet the needs of one will not meet the needs of another. Thus it is clear that the quality of protein can only usefully be described in terms of the amino acids which it supplies to the bird. However, it is still useful to specify total requirements for crude protein in addition to specifying requirements for each essential amino acid since this ensures that the diet supplies sufficient precursors for the non-essential amino acids.

Vitamin requirements

Vitamins are often involved in enzyme systems. Unlike protein, energy and some minerals, they are required by the bird in small quantities. The actual amount required depends on diet, rate of growth or egg production, size of bird and possibly the climate. An imbalance of vitamins can lead to serious disorders. Housed poultry are entirely dependent on the vitamins present in their compounded food so it is important that they are provided in the correct amounts. All vitamins are available in a synthetic form at a commerical price and so the provision of these to poultry is a comparatively simple operation in developed countries. Unfortunately these synthetic vitamins may not be available in tropical countries and without them it is difficult to formulate a balanced diet using tropical foodstuffs.

Vitamin A (retinol or retinal)

In its pure form, vitamin A is found as an alcohol (retinol), an ester, or an aldehyde (retinal). Vitamin A can be provided in the diet either in one of these pure forms or as its **provitamin**, i.e. those carotenes that can be converted to vitamin A in the animal body. Provitamins are present in green vegetable matter and yellow maize and are converted into vitamin A in the wall of the intestine. A good source of carotenes are fresh grains from the oil palm.

> *If the level of vitamin A in the diet is too low, growth rate is below normal and mortality increases. Egg production by adults and fertility of eggs may be reduced. In birds of all ages salivary and tear glands may cease to function.*

Vitamin D (D_2, ergocalciferol and D_3, cholecalciferol)

Vitamin D_3 can be synthesised in the skin of chickens from precursors when the chickens are exposed to sunlight. Thus no problems from vitamin D deficiency should be expected from free-range birds in the tropics. However, when birds are kept under intensive conditions some form of supplementation is essential. This can be provided by the addition of fish oils to the diet, but it is now more usual to provide a dry synthetic preparation. Synthetic vitamin D should be given to poultry in the form of D_3 (cholecalciferol) because they use D_2 (ergocalciferol) inefficiently.

> *If vitamin D_3 is not present in the diet in sufficient quantities, birds are unable to use the calcium and phosphorous in the diet.*

The relation between vitamin D_3, calcium, and phosphorous in the diet is a critical one. Each substance must be included in the diet in the correct amount and in relation to the other. If not, both bone formation and egg shell formation may be adversely affected.

> *Lack of vitamin D_3 in the diet causes rickets in chicks and osteoporosis in adults. The hatchability of eggs produced by hens is reduced and egg production may even cease.*

Vitamin K (K_1, phylloquinone, K_2, menaquinone and K_3, menaphthone)

This vitamin is an essential substance in the mechanism of blood clotting and is therefore often termed the anti-haemorrhagic factor.

If it is deficient, haemorrhages can occur in the legs, breasts and wings and clotting time is delayed.

Vitamin K is present in all green foods and is also readily available in synthetic forms, usually as menadione sodium bisulphate. The needs of birds are small.

Vitamin E (tocopherol)

Vitamin E is a rather unstable substance and is easily destroyed by oxidising agents, particularly by rancid oils. Thus although vitamin E is widely distributed in natural foodstuffs, deficiencies may occur in poultry diets.

Vitamin E deficiency causes a condition in chicks which affects the nervous system called encephalomalacia or 'crazy chick disease'. A deficiency in the diets of breeding hens reduces the hatchability of eggs.

The amount of vitamin E in the natural foods should be discounted. Because it is so easily destroyed, vitamin E should be protected by adding an anti-oxidant, e.g. ethoxyquin, to the food mixture.

Vitamin B complex

The B vitamins are a group of substances which have related functions. The members of the group are thiamin (B_1), riboflavin (B_2), pyridoxine (B_6), pantothenic acid, nicotinic acid, choline, vitamin B_{12}, biotin and folic acid. B vitamins are essential for productivity, especially for maximum growth of chicks.

Vitamin B_1 is needed for the metabolism of carbohydrates, whilst a deficiency of vitamin B_2 can lead to a condition known as curly toe paralysis in chicks. Nicotinic acid is required for normal feathering, and choline is associated with the metabolism of fat. A deficiency of pantothenic acid can lead to dermatitis. Vitamin B_{12} (cyanocobalamin) is required for the development of normal red blood cells. Insufficient biotin has been shown to be associated with a condition of broilers known as fatty liver kidney syndrome (FLKS).

Although the major ingredients of poultry diets supply some B vitamins, the diet of high producing birds needs to be supplemented. Ingredients such as fish-meal, dried distillers solubles and dried skimmed milk are rich in B group vitamins but it is usual to add synthetic B vitamins to the diets of intensively kept flocks.

Mineral requirements

Calcium and phosphorus

The main function of both calcium (Ca) and phosphorus (P) is in production and maintenance of the bones of the body. The skeleton accounts for about 99% of the calcium and 80% of the phosphorus in the body. The two minerals interact with each other both before and after their absorption from the digestive tract. Excessive amounts of either mineral can interfere with the utilisation of the other.

The requirements of poultry for calcium and phosphorus are influenced by the amount of vitamin D in the diet. In general the requirements for calcium and phosphorus increase as the level of vitamin D decreases and vice versa. To a lesser extent supplies of both manganese and zinc also affect the utilisation of calcium and phosphorus. The ratio of Ca:P for growing birds should be within the range 1:1 and 2:1, but a ratio of up to 6:1 is needed for laying birds which require in the region of 4.0 g per day of calcium for egg shell formation.

There is some calcium in all foods, but those of animal origin generally have much higher levels than those derived from plant sources. Phosphorus is found in almost all foods, but much of the phosphorus in seeds such as cereal grains and legumes is in the form of the salts of phytic acid which tend to be unavailable to poultry. For calcium too, availability decreases as phytate level increases. Supplements such as steamed bone flour or bone-meal are added to chick and grower diets to provide extra phosphorus and calcium. For laying birds where the requirement for calcium is 20–30 times greater than for non-layers an additional source is required. This is provided either by oyster-shell grit fed separately or by limestone flour added to the diet.

Requirements of calcium and phosphorus are determined by measuring:

- the rate of growth – the ash percentage of the bones;
- or the composition of bone ash.

It is usually found that more dietary calcium is required for maximum bone ash percentage than for maximum growth.

The requirement of growing pullets for calcium should not be judged simply in terms of that which enables normal growth to be achieved but may also be assessed in terms of subsequent laying performance. For example, it is known that pullets store large amounts of calcium in the medullary bones about two weeks before they commence laying.

The main criteria for determining the requirements of calcium and phosphorus for the laying hen are egg production and shell thickness. It is possible to completely inhibit egg production by feeding diets that are almost completely devoid of calcium. On deficient diets egg production will be reduced although for a short time birds are able to draw on reserves stored in their medullary bones.

Specifications for calcium requirements should be in terms of calcium per day rather than percentage in diets. This consideration is particularly important in the tropics where reduced intake caused by high ambient temperatures may result in too little calcium being consumed each day.

Egg shells normally contain between 1.6 and 2.4 g of calcium. Although skeletal reserves may be used to provide some of this calcium, it is clear that over a period of time virtually all the calcium secreted in the shells must be absorbed from the food if the skeleton is not to be permanently depleted. Under normal conditions less than 50% of the calcium consumed is absorbed, and the maximum amount that can be absorbed in a day is about 2.0 g.

The egg spends twenty hours in the uterus during which time the egg shell is produced. During the first four hours of this twenty-hour period only 9% of the egg shell is secreted, and in the remaining sixteen hours the rate of secretion is much more rapid.

Assuming an average shell calcium content of 2 g, calculations show that during the final hours of shell formation, the rate of calcium deposition exceeds the maximum rate at which calcium can be absorbed in the gut. Therefore some form of skeletal mobilisation must be an essential feature of shell formation.

Manganese

Manganese, as mentioned already, is also concerned with calcium metabolism. An adequate quantity in the diet is essential to promote bone formation and shell strength. For breeding birds, manganese is also required to ensure maximum fertility of eggs.

If manganese is deficient in the diet it is often possible to observe cases of perosis when the achilles tendon slips off its groove behind the hock joint. A deficiency of choline can also give rise to perosis.

Manganese is present in many foods and particularly limestone flour but it is normal to feed a supplement of either manganese sulphate or manganese oxide in the diet.

Iron, copper and cobalt

These elements are essential for the formation of haemoglobin and it is normal to add these minerals in a supplement to poultry diets.

Iodine

A deficiency of iodine will cause a reduction of metabolic rate resulting in lethargy.

It is normally added to the diet in the form of iodised salt.

Salt (sodium chloride)

Salt in the diet improves the utilisation of protein.

Inadequate salt leads to hypertension in young and adult fowls and a cessation of egg production by the latter.

Excess salt results in high levels of water consumption and wet droppings. The recommended concentration of common salt in poultry diets is about 0.5%.

Zinc

Low levels of zinc in the diet can reduce growth rate in young chickens and birds often adopt a 'goose' stepping walk. Feather development may be affected.

Zinc is a constituent of some enzymes, e.g. carbonic anhydrase, and is an activator of others. Diets should contain additional zinc in the form of zinc carbonate unless the feeding containers are galvanised in which case the galvanising will act as a source of zinc.

The calculation of the amount of minerals to be added to poultry diets

The mineral requirements of poultry are defined in terms of the mineral elements although minerals are always added to diets in the form of compounds. It is useful to know what proportion of these compounds are made up of the appropriate element so that the correct amount of this element can be added to the diet. To achieve this objective it is necessary to know the atomic weight of the element and the chemical formula of the compound. The atomic weights of the relevant elements are given in Table 4.4 and the chemical formulae of some of the more commonly used compounds are given in Table 4.5.

Table 4.4 Atomic weights of elements used in poultry diets

Element	Atomic weight
Calcium	40
Copper	63.5
Oxygen	16
Sulphur	32
Phosphorus	31
Iron	56
Hydrogen	1
Manganese	55
Iodine	127
Zinc	65

By combining these two pieces of information it is possible to work out the percentage of the elements in mineral compounds (Table 4.5). Pure

Table 4.5 Chemical formulae of some of the compounds used as mineral supplements for poultry

Compound	Chemical formula (and atomic weight) of constituent elements	% of element in compound
Limestone flour	Ca C O_3 (40) (12) (48)	40% Ca
Zinc carbonate	Zn C O_3 (65) (12) (48)	52% Zn
Anhydrous iron oxide	Fe_2 O_3 (112) (48)	70% Fe
Manganese sulphate	Mn S O_4 (55) (32) (64)	36% Mn
Copper sulphate	Cu S O_4 (63.5) (32) (64)	40% Cu

limestone flour contains 40% calcium, however, commercial samples contain some impurities so the actual calcium content may be as low as 36%.

Formulating poultry diets by hand calculation

Commercial formulation of poultry diets is a complex process which is normally done on a computer. It is possible however to design a simple diet by hand as can be seen from the following example. This example is the calculation of a diet for a laying hen and assumes that only four nutrients need to be taken into account (Table 4.6):

- metabolizable energy;
- crude protein;
- calcium;
- phosphorus.

It is further assumed that only five foodstuffs will be used (Table 4.7):

- maize;
- fish-meal;
- soya bean meal;
- limestone flour;
- bone-meal.

When poultry diets are designed using conventional foodstuffs they may follow a fairly predictable pattern and the approximate quantities of the various ingredients that are found in laying hen and broiler diets are given in Table 4.8.

If the information given in Tables 4.6, 4.7 and 4.8 is used a diet such as that given in Table 4.9 can be postulated. The diet in this table contains 70% maize. Table 4.7 shows that the energy content of maize is 14.5 kJ/g, so that the energy provided by maize in this diet would be:

$$14.5 \times \frac{70}{100} = 10.15 \text{ kJ/g}$$

This calculation is repeated for all the other foodstuffs and the total summated so giving the energy content of the diet (Table 4.9). Similar calculations are made for protein, calcium and phosphorus. Comparison of the nutrient content of this diet with the requirements given in Table 4.6 shows that the diet is too high in energy, too high in protein and too low in calcium. To remedy these imbalances we must make appropriate substitutions. If 10 parts of limestone flour are substituted for 10 parts of maize this will increase the Ca content of the proposed diet by 0.35% and decrease energy by 0.145 kJ/g. If we substitute 10 parts of maize for 10 parts of fish-meal this will decrease the protein level by 0.65−0.08 =

0.57% and increase energy by 0.145−0.13 = 0.015 kJ/g. Therefore to achieve the required level of energy and protein we need to substitute 30 parts of maize for 30 parts of fish-meal (Table 4.10).

The resulting diet, shown in Table 4.10, has a nutrient content which is much closer to the nutrient requirement shown in Table 4.6 but is still too high in energy and too low in calcium. A further improvement would be made if 10 parts per 1000 of calcium carbonate were substituted for 10 parts per 1000 of maize.

Table 4.6 Simplified nutrient requirements for a laying hen

Metabolizable energy (kJ/g)	11.5−12.5
Crude protein %	16.5−17.5
Calcium %	2.5− 3.5
Phosphorus %	0.6− 1.0

Table 4.7 Nutritive value of ingredients to be used in the diet of a laying hen

	Metabolizable energy (kJ/g)	Protein (%)	Calcium (%)	Phosphorous (%)
Maize	14.5	8	0.0	0.4
Fish-meal	13.0	65	5.0	2.5
Soya bean meal	11.0	44	0.25	0.6
Limestone flour	0.6	0	35.0	0.0
Bone-meal	0.0	0	22.0	9.0

Table 4.8 Likely contents of laying hen and broiler diets when conventional foodstuffs are used (kg/tonne)

	Laying hen	Broiler
Cereals	600−700	600−700
Other energy foods	50−200	0− 50
Animal protein	0−100	50−200
Vegetable protein	50−150	50−150
Limestone flour	40− 60	0− 10
Phosphorus source	5− 10	5− 10
Salt	2	2
Vitamin + mineral mix	3	3
Carotenoid pigment	+	−

Table 4.9 First attempt to produce a diet for laying hens based on the information given in Tables 4.7 and 4.8

	Content (kg/tonne)	Metabolizable energy (kJ/g)	Protein (%)	Ca (%)	P (%)
Maize	700	10.15	5.6	0.0	0.28
Fish-meal	100	1.30	6.5	0.50	0.25
Soya bean	150	1.65	6.6	0.04	0.09
Limestone	40	0.00	0.0	1.40	0.00
Bone-meal	10	0.00	0.0	0.22	0.09
	1000	13.10	18.7	2.16	0.71

Table 4.10 Proposed diet for laying hen incorporating the modification suggested in the text

	Content (kg/tonne)	Metabolizable energy (kJ/g)	Protein (%)	Ca (%)	P (%)
Maize	720	10.45	5.76	0.0	0.30
Fish-meal	70	0.91	4.55	0.35	0.18
Soya bean	150	1.65	6.6	0.04	0.09
Limestone	50	0.00	0.0	1.75	0.00
Bone-meal	10	0.00	0.0	0.22	0.09
	1000	13.01	16.91	2.36	0.66

What the above examples show is that even when a very simple diet is being designed with a minimum number of nutrients taken into consideration it is a very long and tedious process to formulate a diet. This is so because substituting one foodstuff for another changes all the nutrients in the diet. For a commercial diet with more than twenty foodstuffs and more than ten nutrients to take into consideration, the possible number of combinations becomes almost infinite. If we then wish to formulate the diet at the lowest possible cost it becomes quite clear that we need to apply computers to this task. If computers are not available the task of formulation can be made somewhat easier by using the Pearson square.

Formulating poultry diets using the Pearson square

If it is desired to achieve a set percentage of a nutrient in a final mix using two ingredients only, a device known as the Pearson square can be used. For example, if it is desired to formulate a diet for laying birds containing 16% protein using two ingredients, maize and soya bean meal, assuming air-dry maize contains 9.0% protein and air-dry soya bean contains 36.0% protein, the desired result is achieved as follows. The protein level required in the complete diet is placed in the middle of the square and the percentage protein content of each food on the two left-hand corners of the square.

To work out the proportion of each foodstuff that is required subtract diagonally across the square the small number from the larger one as shown below. The proportion of each ingredient to achieve a diet containing 16% protein is given at the right-hand side of the square.

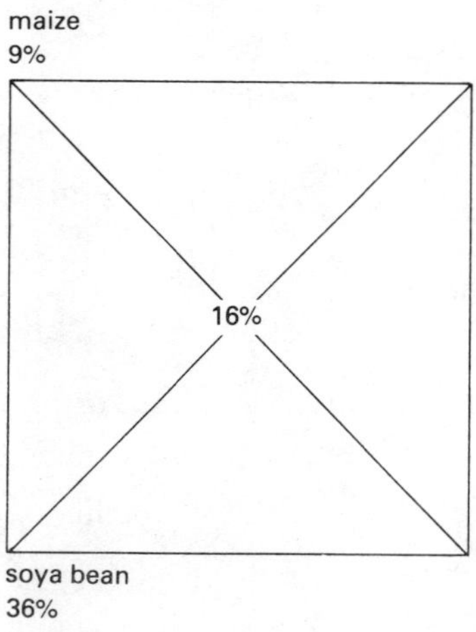

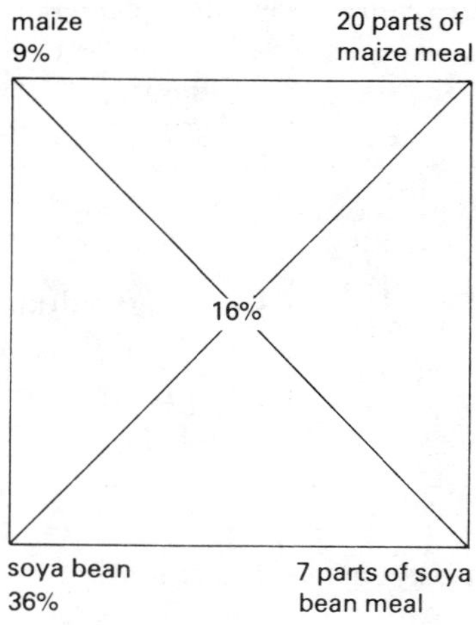

Thus a ratio of 20 parts of maize is required to be mixed with 7 parts of soya bean meal to achieve a diet containing 16% protein. Expressed in percentage terms the diet will be made up as follows:

$$\frac{20}{20 + 7} \times 100\% = 74\% \qquad \text{Maize}$$

and

$$\frac{7}{20 + 7} \times 100\% = 26\% \qquad \text{Soya bean meal}$$

The Pearson square can be used for more than two ingredients provided the various foodstuffs are put into two groups and an average is produced as follows:

Cereals	Maize	9%
	Sorghum	11%
	Average	10%
Protein food	Soya bean meal	36%
	Fish-meal	60%
	Average	48%

These average figures can be used in a Pearson square instead of figures for individual ingredients. The averages could be weighted if different amounts of each ingredient are to be used, i.e. from 2 parts of maize (9% protein) and 1 part of sorghum (11% protein) and 3 parts of soya (36% protein) and 1 part of fish-meal (60% protein).

Cereals – weighted mean

Maize 9 × 2	= 18
Sorghum 11 × 1	= 11
Total	= 29

Weighted mean for cereals $= \frac{29}{3} = 9.7\%$ protein

Protein foods – weighted mean

Soya 36 × 3	= 108
Fish-meal 60 × 1	= 60
Total	= 168

Weighted mean for protein food $= \frac{168}{4} = 42\%$ protein

Thus the average of 9.7% for cereals and 42% for protein foods would be used on the left-hand side of the Pearson square. This technique is useful for calculating one ingredient but not for two or more (e.g. metabolizable energy and protein) and certainly cannot be used to formulate least cost rations.

Other ingredients to be added to poultry diets

Up to this point we have considered only the energy and amino acid content of poultry diets. Other ingredients that should be added include minerals, vitamins, coccidiostats and antioxidants, vitamins and mineral pre-mixes.

A typical mineral pre-mix which could be added to both layer and broiler diets is given in Table 4.11.

A typical vitamin pre-mix which could be added to both diets of laying hens and broilers is given in Table 4.12.

Coccidiostats

A coccidiostat should be introduced at a prophylactic level in every broiler diet. It should be used according to the directions of the manufacturer including its complete withdrawal towards the end of the fattening period. As coccidiosis may occur in flocks treated with a coccidiostat at prophylactic levels it is advisable to have a concentrated coccidiostat available which can be used to medicate the water at a curative level.

Table 4.11 Sample mineral pre-mix for one tonne of broiler or layer diet

Manganese sulphate	332.0 g
Iron sulphate	205.0 g
Zinc carbonate	86.0 g
Copper sulphate	10.0 g
Potassium iodide	1.5 g
Cobalt sulphate	0.5 g
Sodium selenate	0.5 g
Plus inert filler to give a total weight of:	1.0 kg

Table 4.12 Sample vitamin pre-mix for one tonne of broiler or layer diet

Vitamin A (IU*)	6 000 000
Vitamin D_3 (IU)	600 000
Vitamin E (mg)	4 000
Thiamin (mg)	1 500
Riboflavin (mg)	2 500
Niacin (mg)	30 000
Vitamin K (mg)	1 000
Pantothenic acid (mg)	10 000
Folic acid (mg)	500
Vitamin B_{12} (mg)	12
Choline chloride (g)	1 000
Plus inert filler to give a total weight of:	2.0 kg

* International unit

NB Add an antioxidant such as ethoxyquine to food mixture at the rate of 112 g per tonne to prevent oxidisation of vulnerable vitamins such as vitamin E.

Antioxidants
The inclusion of an antioxidant such as ethoxyquine or butylated hydroxytoluene in a food mixture is advisable to prevent the destruction of certain vitamins, especially vitamin E. Vitamin E is particularly vulnerable to oxidation which causes its destruction particularly when unsaturated fat is incorporated in the diet. The inclusion of an antioxidant therefore enables food to be stored safely for a long period of time.

Feeding of turkeys

The carcasses of turkeys contain a higher percentage of protein than the carcasses of chickens. Consequently meat type turkeys have a requirement for a higher protein content in their diets. Details of the nutrient requirements of turkeys are given in Table 4.13.

Feeding of ducks

Unlike chickens, ducks do not have a crop and their proventriculus is cylindrical. The absence of a crop is probably responsible for the faster rate of passage of ingesta in ducks than in chickens. Ducks are best fed food in the form of pellets rather than dry mash as this reduces waste. Waste can also be reduced if the food is fed in the form of a wet mash. Information on the nutrient requirements of ducks is given in Tables 4.14 and 4.15.

Mixing diets for poultry

Vitamins and minerals should always be mixed in separate pre-mixes (usually 15 kg pre-mix per tonne of food). The antibiotics, the coccidiostat and the antioxidant may be pre-mixed with the vitamins (Fig 4.3). Fat should be melted by heating to 71°C and either put into the top of the mixer with a special sprayer or carried into the mixing machine which has bottom force feed auger, using the last quantity of cereals as a carrier.

To ensure that an even mix is obtained the ingredients should be put into the mixer **in the following order**:

1 cereals;
2 vitamin pre-mix (15 kg cereals, plus vitamins, antioxidant and coccidiostat);
3 protein foods;
4 cereals;
5 mineral pre-mix (15 kg cereals plus minerals);
6 cereals.

Table 4.13 Summary of nutrient requirement of turkeys expressed as dietary concentrations (in air-dry material)

Constituent	Minimum			Maximum for all ages
	0–6 weeks	7–12 weeks	over 12 weeks	
Lysine %	1.30	1.04	1.00	–
Methionine %	0.50	0.40	0.36	
Crude fibre %	–	–	–	
Metabolizable energy (kJ/g)	12.60	12.00	12.00	–
Crude protein %	28.00	22.00	18.00	–
Calcium (Ca)%	0.80	0.50	0.50	2.8
Manganese (Mn) (mg/kg)	50.00	50.00	50.00	–
Sodium chloride %	0.50	0.50	0.50	0.5
Phosphorus (P) %	0.65	0.50	0.50	1.0
Zinc (Zn) (mg/kg)	70.00	70.00	70.00	–
Choline (mg/kg)	1 500.00	1 500.00	1 500.00	
Niacin (mg/kg)	35.00	35.00	35.00	–
Pantothenic acid (mg/kg)	10.50	10.50	10.50	–
Riboflavin (mg/kg)	3.50	3.50	3.50	–
Vitamin A_1 (IU*/kg)	1 760.00	1 760.00	1 760.00	–
Vitamin D_3 (IU/kg)	900.00	900.00	900.00	

* International unit

Table 4.14 Summary of nutrient requirements of chicks, broilers and ducklings expressed as dietary concentrations (in air-dry material)

Constituent		Minimum	Maximum
Lysine %	(i) chicks and ducklings	1.0	–
	(ii) broilers	1.1	–
Methionine	(i) chicks and ducklings	0.4	–
	(ii) broilers	0.48	–
Crude fibre %	(i) chicks and ducklings	–	6.0
	(ii) broilers	–	4.0
Metabolizable energy (kJ/g)	(i) chicks	11.0	–
	(ii) broilers and ducklings	13.0	–
Crude protein %	(i) chicks and ducklings (up to 4 weeks)	18.0	–
	(ii) ducklings (4–8 weeks)	16.0	–
	(iii) broilers (up to 5 weeks)	21.0	–
	(iv) broilers (5–10 weeks)	19.0	–
Calcium (Ca) %		0.7	1.3
Manganese (Mn) (mg/kg)		50.0	–
Sodium chloride %		0.4	0.4
Phosphorus (P) %		0.4	1.1
Zinc (Zn) (mg/kg)		40.0	–
Choline (mg/kg)		1 300.0	–
Niacin (mg/kg)		28.0	–
Pathothenic acid (mg/kg)		10.0	–
Riboflavin (mg/kg)		4.0	–
Vitamin A_1 (IU*/kg)		1 320.0	–
Vitamin D_3 (IU/kg)		400.0	–

* International unit   NB Air-dry signifies containing the amount of moisture normally contained in foodstuffs used on the farm.

Table 4.15 Summary of nutrient requirements of laying fowls and laying ducks expressed as dietary concentration (in air-dry material)

Constituent	Minimum	Maximum
Lysine %	0.80	–
Methionine %	0.35	–
Crude fibre %	–	7.0
Metabolizable energy (kJ/g)	12.00	–
Crude protein %: hens	16.50	–
ducks	19.00	
Calcium (Ca) %	3.00	3.5
Manganese (Mn) (mg/kg)	30.00	–
Sodium chloride %	0.50	0.5
Phosphorus (P) %	0.50	1.0
Zinc (Zn) (mg/kg)	60.00	–
Choline (mg/kg)	1 100.00	–
Niacin (mg/kg)	8.00	–
Pantothenic acid (mg/kg)	1.50	–
Riboflavin (mg/kg)	2.50	–
Vitamin A_1 (IU*/kg)	2 700.00	–
Vitamin D_3 (IU/kg)	600.00	–

* International unit

It is generally recognised that small pellets (crumbs) provide the most acceptable type of diet for young poultry and larger pellets for adult birds. They are both readily picked up by the birds and they flow more freely in the feeders, especially tube feeders. Pellets may also contain fewer micro-organisms than meal because they are heated during the preparation process.

Mash is, however, an acceptable substitute especially if it contains a small percentage of fat to reduce dust problems.

Food formulation using a computer

A number of programs are now available which allow diets to be formulated on desk top microcomputers such as the Amstrad 1513.

Fig 4.3 *Preparing a pre-mix for adding to the diet*

Advantages of computer food formulation

The advantages of computer food formulation are as follows.

- It enables a diet to be formulated which, within the constraints given, is the one which meets all the nutritional requirements at the least possible cost.
- It rejects those foodstuffs which are too costly for inclusion in the mix and tells the compounder how much he can afford to pay for the ingredients rejected.
- It informs the nutritionist which of the constraints is raising the price of the mix. It also shows how much the price would be altered if those contraints were changed.
- It gives a complete breakdown of the nutrient content of the ration, thus saving the nutritionist a considerable amount of laborious calculation.
- Compounds can be compiled which do not total 1000 kg but still meet all the nutritional constraints. A filler can be added to this mix to make it up to 4000 kg.

Principles involved in computer food formulation

To use the computer in food formulation it is necessary to have tables of the nutrient contents of foodstuffs. These may not be available in many developing countries so approximate values are given in Tables 4.16–4.20, which should be modified as new information becomes available.

The actual process of food formulation is comparatively simple from the point of view of the nutritionist, provided he can discuss the problem with a computer specialist. The type of program that is needed will be available for most main-frame computer systems and a number have been developed for microcomputers (Fig 4.4).

- The first job is to enter into the computer all the information about the foodstuffs, such as that given in Tables 4.16–4.20. This information can remain permanently in the computer memory or on floppy disc.
- The second task is to enter the price per tonne of all the ingredients to be considered by the computer.
- The third task is to specify the nutrient requirements of the birds. These specifications are given in Tables 4.13, 4.14, and 4.15.
- The fourth task is to specify limitations to be placed on the usage of different types of foodstuff (Table 4.21).

These limitations may be of a nutritional nature, for example the nutritionist may decide to limit the amount of blood-meal used in any diet. Certain foodstuffs which are cheap may only be available in limited quantities and therefore he/she may wish to use them sparingly in certain types of diet.

Fig 4.4 *Students at the Centre for Tropical Veterinary Medicine in Edinburgh learning to use microcomputers in the formulation of diets*

Table 4.16 Nutrient content of vegetable protein foodstuffs used in poultry foods in the tropics and subtropics (in air-dry material)

	Cottonseed cake meal		Groundnut cake meal		Lucerne meal	Soya bean	Soya bean (full fat)	Sunflower cake meal
	1	2	1	2				
Lysine %	1.60	1.40	1.80	2.20	1.00	2.90	2.40	1.50
Methionine %	0.52	0.43	0.15	0.43	0.15	0.60	0.51	0.80
Crude fibre %	14.00	18.00	5.00	5.00	30.00	6.00	4.00	12.00
Metabolizable energy (kJ/g)	9.68	9.32	10.67	10.75	2.90	11.17	14.50	11.79
Crude protein % (N × 6.25)	40.00	36.00	48.00	52.00	16.00	44.00	37.00	40.00
Calcium (Ca) %	0.20	0.20	0.16	0.16	1.40	0.25	0.22	0.41
Manganese (Mn) (mg/kg)	20.00	20.00	18.00	18.00	50.00	30.00	26.00	23.00
Sodium chloride %	0.50	0.50	0.12	0.12	0.50	0.34	0.29	–
Total phosphorus (P) %	1.20	1.20	0.80	0.80	0.30	0.60	0.52	0.95
Zinc (Zn) (mg/kg)	–	–	–	–	35.00	28.00	24.00	–
Choline (mg/kg)	2 300	2 000	2 000	2 000	800	2 000	1 730	280
Niacin (mg/kg)	40	40	200	200	25	20	17	250
Pantothenic acid (mg/kg)	10	10	30	30	14	12	23	38
Riboflavin (mg/kg)	5	5	5	5	13	26	22	3
Vitamin A_1 (IU*/kg)	300	300	330	330	20 000	340	290	–
Vitamin D_3 (IU/kg)	–	–	–	–	–	–	–	–

* International unit

Table 4.17 Nutrient content of animal protein foodstuffs used in poultry foods in the tropics and subtropics (in air-dry material)

	Blood-meal	Fish-meal	Meat and bone-meal	Skim milk (dried)	Whale meal
Lysine** %	7.00	5.00	3.00	2.50	4.00
Methionine** %	0.90	1.80	0.50	0.80	1.00
Crude fibre %	1.00	1.00	1.00	–	1.00
Metabolizable energy (kJ/g)	11.77	12.60	10.11	11.12	12.09
Crude protein % (N × 6.25)	80.00	65.00	48.00	33.00	80.00
Calcium (Ca) %	0.28	5.00	12.50	1.30	2.00
Manganese (Mn) (mg/kg)	5.00	40.00	12.00	2.00	290.00
Sodium chloride %	1.50	2.50	1.75	1.65	0.50
Total phosphorus (P) %	0.22	2.70	5.50	1.00	0.50
Zinc (Zn) (mg/kg)	–	68.00	100.00	–	
Choline (mg/kg)	750.0	3 000.0	1 500.0	1 100.0	2 200.0
Niacin (mg/kg)	30.0	60.0	40.0	11.0	66.0
Pantothenic acid (mg/kg)	1.1	8.0	3.0	3.0	3.0
Riboflavin (mg/kg)	1.5	6.0	3.5	20.0	2.2
Vitamin A_1 (IU*/kg)	–	–	–	260.0	–
Vitamin D_3 (IU/kg)	–	–	–	–	–

* International unit

** Amino acids vary according to the protein level

Table 4.18 Nutrient content of cereal grains used in poultry foods in the tropics and subtropics (in air-dry material)

	Corn and cob	Maize (white)	Maize (yellow)	Munga	Sorghum
Lysine** %	0.20	0.24	0.27	0.20	0.22
Methionine** %	0.12	0.14	0.16	0.27	0.18
Crude fibre %	6.00	2.00	2.00	2.00	2.00
Metabolizable energy (kJ/g)	–	14.69	14.57	13.92	14.48
Crude protein % (N × 6.25)	7.60	8.00	9.00	10.50	10.00
Calcium (Ca) %	–	–	–	0.05	0.05
Manganese (Mn) (mg/kg)	13.00	5.00	5.00	5.00	13.00
Sodium chloride %	0.07	0.08	0.08	0.05	0.13
Total phosphorus (P) %	0.35	0.40	0.40	0.40	0.30
Zinc (Zn) (mg/kg)	20.00	23.00	23.00	20.00	15.00
Choline (mg/kg)	360.00	440.00	440.00	440.00	700.00
Niacin (mg/kg)	14.00	16.00	16.00	16.00	32.00
Pantothenic acid (mg/kg)	3.00	3.30	3.30	3.30	9.00
Riboflavin (mg/kg)	0.60	0.66	0.66	0.50	1.00
Vitamin A (IU*/kg)	–	–	4 400.00	–	–
Vitamin D_3 (IU/kg)	–	–	–	–	–

* International unit
** Amino acids vary according to the protein level
NB Values obtained by methods of chemical analysis shall take precedence over the average figures in the table above.

Table 4.19 Nutrient content of other energy foodstuffs used in poultry foods in the tropics and subtropics (in air-dry material)

	Cane molasses	Hominy chop	Maize bran	Maize germ	Sugar	Wheat bran	Wheat pollard
Lysine** %	–	0.30	0.30	0.35	–	0.60	0.75
Methionine** %	–	0.08	0.08	0.14	–	0.30	0.20
Crude fibre %	–	6.00	13.00	5.00	–	12.00	10.00
Metabolizable energy (kJ/g)	10.16	7.20	6.38	14.69	17.52	7.02	12.85
Crude protein %	3.00	9.00	7.00	10.00	–	14.00	16.00
Calcium (Ca) %	0.70	0.05	0.03	0.04	–	0.10	0.10
Manganese (Mn) (mg/kg)	42.00	7.00	7.00	13.00	–	120.00	110.00
Sodium chloride %	–	–	–	0.25	–	0.12	0.05
Total phosphorus (P) %	0.10	0.50	0.50	0.50	–	1.10	1.10
Zinc (Zn) (mg/kg)	–	23.00	23.00	20.00	–	130.00	130.00
Choline (mg/kg)	850.0	440.0	440.0	1 500.0	–	1 000.0	1 000.0
Niacin (mg/kg)	33.0	50.0	40.0	40.0	–	200.0	90.0
Pantothenic acid (mg/kg)	38.0	7.0	5.0	4.0	–	29.0	16.0
Riboflavin (mg/kg)	3.3	2.0	1.5	3.0	–	3.0	1.6
Vitamin A (IU*/kg)	–	–	–	–	–	1 000.0	1 000.0
Vitamin D_3 (IU/kg)	–	–	–	–	–	–	–

* International unit

** Amino acids vary according to the protein level

NB Values obtained by methods of chemical analysis shall take precedence over the average figures given in the table above.

Table 4.20 Nutrient content of substances used as mineral supplements in poultry foods in the tropics and subtropics (in air-dry material)

	Bone-meal	Dicalcium phosphate	Limestone flour	Monocalcium phosphate	Monosodium phosphate
Crude fibre %	2.0	–	–	–	–
Metabolizable energy (kJ/g)	5.2	–	–	–	–
Crude protein % (N×6.25)	25.0	–	–	–	–
Calcium (Ca) %	22.00	22	37.00	16	–
Manganese (Mn) (mg/kg)	30.00	700	1 000.00	–	–
Sodium chloride %	0.80	–	0.10	–	–
Total phosphorus (P) %	9.00	18	0.04	20	22
Zinc (Zn) (mg/kg)	425.00	–	–	–	–
Niacin (mg/kg)	4.0	–	–	–	–
Pantothenic acid (mg/kg)	2.0	–	–	–	–
Riboflavin (mg/kg)	0.9	–	–	–	–

NB Values obtained by methods of chemical analysis shall take precedence over the average figures given in the table above.

Table 4.21 Specifications for a broiler diet and the results achieved when the information was fed into a least cost diet program. The last column shows the effect of the constraints applied on the cost of the diet

Class of foodstuff	Foodstuff & nutrients	Constraints applied		Actual ration	Affect of constraint on unit price of diet.*
Cereals	Total	=	1 000		
	Maize	Not >**	750	52.53	
	Millet	Not >	250	250.00	102.1
	Sorghum	Not >	100	100.00	13.2
By-product	Sugar	Not >	100	100.00	61.2
	Maize bran	Not >	150	–	
	Pollard	Not >	300	–	
	Wheat bran	Not >	100	64.27	
	Lucerne meal	Not >	50	–	
	Maize germ	Not >	50	50.00	55.8
Vegetable protein	Cottonseed	Not >	50	50.00	13.2
	Peanut meal	Not >	300	182.86	
Animal protein	Blood-meal	Not >	25	25.00	279.6
	Meat & bone-meal	Not >	50	50.00	105.6
	Fish-meal	Not <***	50	58.55	
Minerals	Bone-meal	Not >	30	–	
	Monocalcium phosphate	Not >	40	2.65	
	Limestone flour	Not >	40	9.61	
	Salt	=	4	4.00	
Amino acids	Methionine	Not >	10	0.53	
	Lysine	Not >	10	–	
Nutrients	Protein %	Not <	22		
		Not >	24	24.00	240.0
	Metabolizable energy % (kJ/g)	Not <	12.5	12.50	1 008.0
		Not >	13.5	12.50	10.8
	Fibre %	Not <	2	3.49	
		Not >	5		
	Calcium %	Not <	0.8	1.30	
		Not >	1.3		37.2
	Phosphorus %	Not <	0.8	0.80	2 076.0
		Not >	1.2		
	Lysine %	Not <	1.15	1.15	8 526.0
	Methionine & cysteine %	Not <	0.75	0.75	10 378.8
	Tryptophan %	Not <	0.20	0.29	

* Relaxing the constraint by one unit would decrease price of diet by amount shown
** > more than
*** < less than

Table 4.22 Changes in the composition of the diet as a result of changes in the constraints on individual foodstuffs

Class of foodstuff	Foodstuff & nutrients	Control	No limit cotton & blood	No limit calcium	Total equal 1 000
	Total	1 000.00	1 000.00	1 000.00	956.22
	Price (currency units)	2 445.00	2 411.00	2 443.00	2 430.00
Cereals	Maize	52.53	134.48	73.55	68.92
	Millet	250.00	250.00	250.00	250.00
	Sorghum	100.00	100.00	100.00	100.00
By-products	Sugar	100.00	100.00	100.00	100.00
	Maize bran	–	46.76	–	–
	Pollard	–	–	–	–
	Wheat bran	64.27	–	–	–
	Lucerne meal	–	–	–	–
	Maize germ	50.00	50.00	50.00	50.00
Vegetable protein	Cottonseed	50.00	135.69	50.00	–
	Peanut-meal	182.86	–	197.61	239.62
Animal protein	Blood-meal	25.00	58.00	25.00	25.00
	Meat and bone-meal	50.00	50.00	50.00	50.00
	Fish-meal	58.55	50.00	60.72	63.77
Minerals	Bone-meal	–	20.24	–	–
	Monocalcium phosphate	2.65	–	5.29	4.19
	Limestone flour	9.61	–	33.35	–
	Salt	4.00	4.00	4.00	4.00
Amino acids	Methionine	0.53	0.83	0.48	0.73
	Lysine	–	–	–	–
Nutrients	Protein %	24.00	22.00	24.00	24.00
	Metabolizable energy (kJ/g)	12.50	12.50	12.50	12.50
	Fibre %	3.49	3.44	2.78	2.43
	Calcium %	1.30	1.27	2.19	0.99
	Phosphorus %	0.80	0.80	0.80	0.80
	Lysine %	1.15	1.15	1.15	1.15
	Methionine & cysteine %	0.75	0.75	0.75	0.75
	Tryptophan %	0.29	0.29	0.28	0.27

Specimen diets produced by the computer

As an example to show how a computer is used, four diets for broilers have been formulated on the computer using ingredients available in Central Africa. The results are shown in a series of tables. Table 4.21 is the most important one. It lists all the ingredients used, the nutrient specification, and the diets formulated.

In the upper part of the first column of this table the types of foodstuffs are listed. The ingredients used have been divided up into classes according to their nutrient contents, i.e. cereals, by-products, vegetable and animal proteins, minerals, and synthetic amino acids. The next three columns above the horizontal line show the actual foodstuffs made available to the computer, the constraints applied to these ingredients and the ration calculated respectively. Below the horizontal line are the nutrients taken into consideration (column 2), the levels specified (column 3), the levels of nutrient in the ration formulated by the computer (column 4), and the effects on the constraints applied on price (column 5).

According to the specifications given, the total weight of the diet had to equal 1000 kg. The computer was allowed to select up to 750 kg of maize but it selected only 52.53 kg. Millet was restricted to a maximum of 250 kg. With the given cereal price structure, the computer selected millet in preference to sorghum, and sorghum in preference to maize. In order to make the computer select all three cereals, it was necessary to place upper limits on the use of both millet and sorghum. As can be seen, the computer selected both millet and sorghum at the upper limits. Both of these constraints increased the price of the ration, and if the constraint on millet usage had been reduced from 250 to 249 kg, then the price of the diet would have been decreased by 102 units[1] per tonne (column 5).

Likewise, if the sorghum constraint had been raised from 100 to 101 kg, the price would have decreased by 13.2 units per tonne. If these constraints are tightened, i.e. changes from 250 to 249 kg and 100 to 99 kg respectively the price of the diet would be increased by similar amounts.

The other constraints which influenced the final price of the ration can be identified from column 5. These are the upper limits placed on sugar, maize germ, cottonseed, blood- and meat- and bone-meal. The values listed in column 5 only apply over a narrow range of change, but they are very useful to the nutritionist in that they show him which constraints to relax if he wishes to reduce the price of the diet.

1 Units rather than a currency are quoted so that the principle can be understood in any country.

Only those nutrients that are liable to be deficient or in excess were specified. No account is taken of the vitamins and trace minerals, because a mineral-vitamin supplement can easily be added to the final diet. To include these as constraints would result in little saving in the final cost of the diet but would add to the time and therefore to the cost of computer formulation.

In this example the amount of protein allowed in the diet protein was given both upper and lower constraints, and, rather unexpectedly, the upper level 24%, was selected. Should this be relaxed by 1% the price would go down by 240 units per tonne. The reason for this apparent anomaly is that the amino acid methionine could have been provided more cheaply from animal protein sources rather than synthetic methionine if the upper protein limitation had been relaxed. The need to include lysine and methionine (column 5) plus cysteine had the greatest effect on the price of the final diet.

Maize bran and bone-meal were not selected. This indicating that they were offered at too high a price. The computer can be used to ascertain at which price they would have been included. This information is a very valuable management aid as it tells the nutritionist the price he should pay for alternative foodstuffs.

Effects of various constraints applied when formulating a diet

Constraints on individual foodstuffs

The main problem when considering constraints on individual foodstuffs is to achieve a compromise between using too few constraints and too many. With too few constraints a diet might be produced, which, although meeting the stated nutritional requirements, would not be nutritionally acceptable. If on the other hand, too many constraints are applied, the diet might just as well have been compiled by hand as the computer is given too little scope. Too many constraints can also lead to the computer being unable to formulate a diet because the specification cannot be met.

In the next example (Table 4.22) no upper limits have been placed on the amounts of cottonseed and blood-meal. This has had the effect of cheapening the diet by 34 units per tonne. The cottonseed and blood-meal levels in the ration are, however, nutritionally unacceptable. In addition, it can be seen that by changing these two constraints the whole structure of the diet has been changed, i.e. the amount of synthetic methionine has increased to 0.83 kg due to a decrease in the fish-meal content while the total protein has dropped to 22%.

Constraints applied to individual nutrient requirements

Consideration is given, for example, to limits on individual amino acids (particularly, lysine, methionine and tryptophan) total protein, metabolizable energy, fibre and the minerals, calcium and phosphorus. Maxima and minima may need to be specified, particularly in the case of cheap ingredients such as calcium. If not, the machine would select far too much of the cheap ingredients.

In the example given in Table 4.22 an upper limit for calcium was not specified and the computer selected a level of 2.19% calcium in the final diet. This proportion would be nutritionally undesirable.

Weight of the final mix

Usually diets are computed in mixes of 1000 kg but it is possible to provide nutrients of particular stock within a smaller mix than this. If a cheap filler could be found, it would be economic to produce a diet and make it up to 1000 kg with the filler.

In the example given in Table 4.22 a diet was formulated with a total weight of less than 1000 kg. The total weight of the diet formulated was 956.22 kg. The cost was 24.31 units fewer per tonne than the control, assuming a filler could be provided at no cost. Thus, the specified nutrients would be supplied at a lower cost than in the example of least cost ration where a 1000 kg mix was specified.

Quality of the final diet produced

The next decision that has to be taken is – are we interested in obtaining the cheapest ration per tonne, or the one which will give the greatest productivity per unit cost? A high energy ration will normally give better food conversion ratios than a low energy ration, and within limits a high protein diet will be better than a low protein diet. In general we are interested in producing not the cheapest ration but the one which will give the greatest return per unit spent on food. This is not easy to achieve as individual strains of animals respond differently to different rations.

The way in which the price of the ration is changed by varying energy levels is illustrated in Table 4.23 and Fig 4.5. The increase in price is not the same for each change and is especially severe between 13.5 and 14 kJ/g (Fig 4.5). The reason for this is the inclusion of fat in the high energy diets which is a very expensive source of energy.

The calcium content of the diet decreased as the energy content increased due to the diet becoming more concentrated and calcium being removed to make room for energy rich foodstuffs. Note also the decrease

Table 4.23 Changes in diet composition and price as a result of changing the energy specification of the diet

Class of foodstuff	Foodstuff & nutrients	Control	Metabolizable energy (kJ/kg)				
			12.00	12.50	13.00	13.50	14.00
	Price (currency units)	2 445.00	2 397.00	2 445.00	2 492.00	2 586.00	2 782.00
Cereals	Maize	52.53	1.35	64.16	108.00	217.40	253.22
	Millet	250.00	250.00	250.00	250.00	250.00	250.00
	Sorghum	100.00	100.00	100.00	100.00	67.69	–
By-products	Sugar	100.00	100.00	100.00	100.00	100.00	100.00
	Maize bran	–	–	–	–	–	–
	Pollard	–	6.86	–	–	–	–
	Wheat bran	64.27	100.00	28.76	–	–	–
	Lucerne meal	–	–	–	–	–	–
	Maize germ	50.00	50.00	50.00	50.00	50.00	50.00
	Peanut oil	–	–	–	–	–	–
	Lard	–	–	–	–	–	23.34
Vegetable protein	Cottonseed	50.00	50.00	50.00	29.36	–	–
	Peanut meal	182.86	180.45	191.00	207.50	136.18	144.31
Animal protein	Blood-meal	25.00	25.00	25.00	25.00	25.00	25.00
	Meat & bone-meal	50.00	50.00	50.00	50.00	50.00	50.00
	Fish-meal	58.55	56.42	59.75	62.73	90.47	90.51

Table 4.23 (cont)

Class of foodstuff	Foodstuff & nutrients	Control	Metabolizable energy (kJ/kg)				
			12.00	12.50	13.00	13.50	14.00
Minerals	Bone-meal	–	–	–	–	3.98	4.75
	Monocalcium phosphate	2.65	1.22	4.09	4.46	–	–
	Limestone flour	9.61	24.10	22.72	8.37	4.45	4.03
	Salt	4.00	4.00	4.00	4.00	4.00	4.00
Amino acids	Methionine	0.53	0.59	0.51	0.55	0.81	0.84
	Lysine	–	–	–	–	–	–
Nutrients	Protein %	24.00	24.00	24.00	24.00	22.00	22.00
	Metabolizable energy (kJ/g)	12.00	12.00	12.50	13.00	13.50	14.00
	Fibre %	3.49	3.90	3.10	2.67	2.17	2.15
	Calcium %	1.30	1.30	1.30	0.80	0.80	0.80
	Phosphorus %	0.80	0.80	0.80	0.80	0.80	0.80
	Lysine %	1.15	1.15	1.15	1.15	1.15	1.15
	Methionine & Cysteine %	0.75	0.75	0.75	0.75	0.75	0.75
	Tryptophan %	0.29	0.29	0.29	0.29	0.25	0.25

in the level of low energy foodstuffs and the increase in ingredients of animal origin. The protein content of the ration also drops from 24% to 22% and in the last two diets bone-meal replaces monocalcium phosphate possibly because of the protein and higher calcium content of the former ingredient. The cost per unit of energy is lowest at 13 kJ/g and increases as the energy content of the ration is decreased or increased from this point (Fig 4.5).

Rounding off

Millers are compelled for technical reasons to weigh in discrete units whereas the computer gives answers to several decimal places. Rounding off can be achieved by the computer which normally leads to a slight increase in the price of the total diet or to a slight decrease in diet quality.

Foodstuffs commonly used in poultry diets in the tropics

Food represents approximately 60% of the cost of producing table poultry or eggs for human consumption. Savings in the use of poultry food can therefore make a major contribution to the economics of production. It is therefore important that the cheapest suitable food is used in the correct

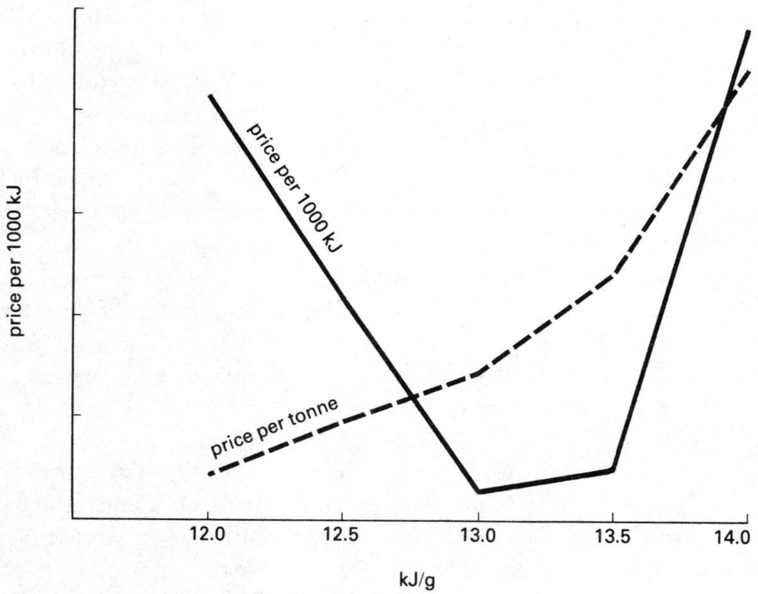

Fig 4.5 *The effect of increasing the energy content of a broiler ration on price per tonne of ration and price per thousand kilojoules of energy*

ratio in poultry diets. It is also important to know what foods are available and how they should be used in poultry diets. Foods used in poultry diets can be classified into five broad classes:

- cereals;
- other energy foods;
- animal proteins;
- plant proteins;
- mineral supplements.

Cereals

Most of the energy in cereal grains is present in the form of starch. Of the cereals, wheat and grain sorghum have the highest content of metabolizable energy. Oats and maize have a high oil content whereas the oil content of wheat, barley, sorghum and millet is low.

Although cereals contain relatively low percentages of crude protein (8–12% of the air-dried food) they are used in high percentages in the diets of poultry so they contribute from one-third to one-half of the protein in the diet of chickens. The amino acid balance in cereals is very different from that found in poultry meat or eggs, and they have a low biological value (50–60) (page 96). If an adequate diet is to be provided for poultry cereals need to be supplemented not only with foodstuffs containing higher protein levels than they contain but also with foodstuffs that complement them for amino acid content.

With regard to vitamins, none of the grains, with the exception of yellow maize, contain important amounts of carotene. All grain contains satisfactory amounts of vitamin E and thiamine but contains inadequate amounts of riboflavin for poultry nutrition. Of the major minerals, grain contains moderate amounts of phosphorus, although a considerable quantity may be locked in the form of phytic phosphorus and thus unavailable to poultry. Grains are very low in calcium.

In countries with a dry climate, the dry matter content of cereals is in the region of 90%, whilst in wetter climates it may range from 75–85%. For prolonged storage the level of moisture in the grain should not exceed 14%, therefore in wetter climates grains often have to be dried before they can be stored.

Maize or corn (*Zea mays*)

This grain which is widely grown throughout the tropics and subtropics is used extensively in diets for poultry. It has a high energy content but a comparatively low protein content. The protein in maize is of a poor quality being deficient in both tryptophan and lysine. Most hybrid varieties tend to produce grain of a lower protein content than do the old open-pollinated varieties. Yellow maize has a protein content of 10%

and white maize 8%. Success has been achieved in increasing the protein (to 11–15%), lysine and tryptophan content of maize by transferring the gene 'opaque 2' and 'floury 2' into various varieties. Unfortunately these modified varieties have a lower yield (by 10%) than the original, reducing the benefit of the increased protein yield. Yellow maize is advantageous where yellow fleshed broilers are required and the yellow pigmentation also helps in the colouring of egg yolks. However in countries where white fleshed broilers are preferred, white maize or other non-colouring cereals must be used because the broilers currently in use have the genetic properties to lay down yellow coloured fat.

Maize is an excellent food for both meat and egg laying birds and is often the basis of poultry diets in developing countries.

Sorghum, milo or kaffir corn (*Sorghum vulgare*)

This grain is grown in drier regions of the tropics than maize. It tends to be lower in energy but higher in protein than maize. Some varieties of sorghum have a high tannin content which renders them less palatable and therefore less useful in poultry diets. Varieties which are low in tannin are available. Dwarf varieties of sorghum are widespread in the USA, these are high yielding but grow only to a height of 2–4 feet. The grain from sorghum resembles maize in composition and feeding value. Its energy content is about 5–10% lower than that of maize but its protein content is 10–12% higher. The seeds are of a small size which can lead to some difficulties with handling and grinding.

Bullrush millet (*Pennisetum typhoides*)

This cereal is widely grown in the drier areas of the tropics. It has very small grains and, like sorghum, is susceptible to predators like the quelea bird (*Quelea quelea*). It resembles maize in composition and feeding value but is considerably higher in protein value. When ground, it has been shown to be an excellent constituent of poultry diets. It has a similar feeding value to barley and like barley promotes the production of white fat.

This cereal is very prone to infection by the fungus *Claviceps purpurea* in the dry season which can produce toxins that are dangerous to both humans and poultry.

Finger millet (*Eleusine coracana*)

This grain is also grown in the drier areas of the tropics although it will grow in higher rainfall areas.

It has the advantage that the grain will not shatter if the crop is left unharvested after the rains. It has a lower protein but slightly higher fibre content than bullrush millet. It can be used in diets for both growing and laying poultry.

Wheat (*Triticum aestivum*)

This crop is seldom grown in the tropics except under irrigation in the dry season in savannah areas and at high altitudes. It is, however, common in Mediterranean countries and is often imported into the tropics from the temperate regions.

It is mainly used for human consumption, particularly for bread making. It has a slightly lower energy content than maize but a higher protein content. Hard wheat has an average protein content of 13–15% whilst soft wheat is much lower with an average content of 10%. Wheat is a very glutinous food and if fed in a ground condition should be limited to 30% of the diet. Higher levels are possible if the wheat is fed whole.

Barley (*Hordeum vulgare*)

Rarely grown in the tropics but commonly grown in Mediterranean regions.

Its protein level is similar to that of soft wheat but it has a much higher fibre content (5%) which makes it an unsuitable ingredient for inclusion in diets for highly productive laying birds and broilers, especially under hot climatic conditions. It could however be a useful ingredient in the diet of broiler breeders where the desire is to reduce food intake.

Rice (*Oryza sativa*)

Threshed rice has a thick, fibrous husk which contains up to 20% silica and in this form it is unpalatable to poultry. The outer casing is easily removed and a much more palatable product, brown rice, is produced. This food is high in starch and low in oil and is a valuable poultry food.

Cereal by-products

Hominy chop or hominy feed

This is the by-product of the manufacture of maize. It is a mixture of maize bran, maize germ and parts of the kernels. It is slightly lower in energy than maize but is higher in protein and fibre.

Maize bran

Maize bran consists of the outer coating of the maize kernel. It contains 15% fibre and is low in metabolizable energy. It is not a suitable ingredient for inclusion in the diet of high producing broilers or laying hens.

Wheat bran

This foodstuff consists almost entirely of the coarse outer coating of the wheat kernel. It contains about 10% fibre which limits its value for poultry food but the protein of the bran is of a better quality than that of wheat or maize. It also has a phosphorous content which is one of the highest of all the common foods.

Rice bran

This is a by-product arising from the milling of rice. It is high in fibre and also in oil. The latter constituent may cause problems of rancidity in food during storage. It probably should not be added to poultry diets at levels above 20% of the total ration.

Wheat pollard or middlings

This food consists of fine wheat bran mixed in varying proportions with wheat-meal. A good quality pollard should consist of 50% bran and 50% wheat-meal. It is appreciably more digestible than wheat bran and has a lower fibre content. It has a relatively high protein content of 14–16%.

Brewers dried grains

Normally a by-product of barley but those from sorghum have a very similar composition. The protein content is in the region of 20% and the fibre content 14%. This high fibre content means that they are of limited value for poultry.

Dried brewer's yeast

This is produced by drying waste yeast from breweries. It contains practically no fibre or oil and the ash is very rich in phosphate. It also contains 40–45% protein of a very high biological value. It is very rich in vitamin B, and riboflavin but is deficient in vitamin B_{12}.

Other energy foods

Cassava, tapioca or yuca (*Manihot esculenta*)

Cassava is a crop that will grow well under poor soil conditions and will also produce very high yields of up to 75 tonnes per hectare under good conditions. It is not recommended that fresh cassava be fed to poultry because of its low dry matter content and the presence of cyanogenetic glucosides. The intact tubers of cassava contain little free cyanide but any cutting or bruising releases free cyanide. If the tubers are chopped into thin slices and left in the sun and allowed to dry to a moisture content below 14% (from 65%) the cyanide content is reduced to an acceptable level. This dried cassava can then be milled and fed to poultry. The major problems are that it has a low protein content and a balance of amino acids very different to that required for poultry. In

West Africa certain types of cassava contain virtually no protein. This lack of protein can cause serious nutritional problems in poultry if adequate amounts of high protein foods are not added to the diet.

Sugar (Sucrose)
Crystalline sugar contains no protein and its value to poultry as a food is lessened by this fact. However, there would appear to be no limit to the amount of sugar that can be added to poultry rations, provided the diets are balanced. The expansion of the sugar cane crop in West Africa could lead to this ingredient becoming very important in poultry diets in this region.

Final molasses
This is the residue left once raw sugar has been extracted from clarified sugar cane juice. It contains sugars and 10–12% soluble mineral matter. It is unsuitable for feeding to poultry because it leads to diarrhoea and wet droppings and also sticks to the feathers of poultry.

Oil seeds and their by-products

The primary reason for growing oil seeds is for the edible oil that they contain. The oil seed meal is the residue remaining after the greater part of the oil has been removed from the seed. These residues are rich in protein and are valuable food for including in poultry diets.

The old method of removing oil by hydraulic pressure is now obsolete. In modern plants either an expeller screw press or an organic solvent is used to extract the oil. In the expeller process the prepared seed is forced by a powerful screw along a narrowing tube so that the oil is squeezed out. The residue is left in the form of broken flakes and normally has an oil content of 5–6%.

In the oil extraction process a non-inflammable volatile solvent such as trichlorethylene or acetone is allowed to percolate through the coarsely ground seed and most of the oil is dissolved out and only 1–2% will be left in the cake.

Sometimes the process is more complex in that the husk or hull of the seed has to be removed before extracting the oil, this leaves a residue after extraction, termed decorticated cake or meal.

It can be seen that a number of different types of oil seed meal are produced:
- expelled;
- extracted;
- undecorticated;
- decorticated.

Each of these will have a different feeding value.

Soya bean (*Glycine max*)

Apart from animal protein foods, soya beans are the source of the protein most suitable for poultry. In their unextracted state they are high in fat (18%), low in fibre (5%), and have 38% protein. Soya beans are rather low in calcium (0.25%) and in phosphorous (0.60%).

The nutritive value of soya beans and soya bean oil meal for poultry is increased by cooking. Cooking greatly increases the availability and the value of the protein for poultry and also destroys the trypsin inhibitor contained within the seed. This substance depresses the growth of non-ruminants and prevents the action of the protein digestive enzymes. Raw soya bean protein has therefore a low feeding value for poultry, but properly cooked whole soya bean or soya bean oil meal provides protein which is nearly equal in value to the protein of meat or fish-meal. Too high a temperature or prolonged cooking reduces the value, because it destroys or makes unavailable lysine and certain other amino acids. The correct temperature regime is 105°C for sixty to ninety minutes. The cooking of soya bean can be achieved on farms. The protein of soya bean is largely in the form of sulphur-containing amino acids – methionine and cysteine.

When the oil has been extracted from the seeds the protein level increases to 40–45% and then decreases to 1.5% if solvent extraction using hexane is carried out.

Groundnut meal (*Arachis hypogaea*)

Whole groundnuts, including the shells, have an average content of 25% protein, they are high in fibre and have a very high metabolizable energy content because of their high oil content of 36%. It is the decorticated groundnut meal which is normally fed to poultry. This product has a fibre content of 6–10% and a protein level of 40–48%. The quality of the protein of groundnut meal is good, ranking close to that of soya bean meal. Its one major disadvantage is that meal is often contaminated with aflatoxin which renders it a very suspect poultry food. Turkeys and ducks are particularly vulnerable to aflatoxins and should not be fed on diets containing groundnut meal.

Cottonseed meal (*Gossypium* spp.)

Better quality cottonseed meal consists of the cake after the oil has been extracted and from which the lint has been removed. The quality will vary according to the amount of lint left in the seed. A good quality meal is light yellow and has a sharp nutty smell. A dark colour indicates that it contains a high percentage of shells and fibre.

Most of the cottonseed meal, including the local product, has 5% of fat or more. As a protein source for poultry it is low in lysine. Cottonseed meal is one of the richest foods in phosphorous, containing 1% or more of this important mineral. In contrast, it has only about 0.2% calcium.

For poultry the ordinary kind of cottonseed meal should be fed in strictly limited amounts because of the danger of injury from a substance called gossypol which cottonseed meal contains. Raw cottonseed contains appreciable amounts of gossypol, the amount varying considerably with the climate and the soil. Gossypol is poisonous to poultry if consumed in sufficient amounts.

Fortunately, in the process of oil manufacture the heat changes much of the gossypol into 'bound gossypol' which is much less poisonous. Nevertheless, it is not safe to include more than 10% cottonseed meal in the ration of poultry. The symptoms of gossypol poisoning are:

- anaemia;
- diarrhoea;
- eventual paralysis.

If the ration of laying hens has more than about 5% of ordinary cottonseed meal, the yolks of the eggs are apt to develop an olive green or brown colour and the white, a pinkish colour on storage. Commercial extraction processes are available which will remove gossypol from cottonseed meal by extraction with the solvent acetone.

Sunflower seed (*Helianthus annus* L.)

The seed contains 25–26% of oil and so its use as a poultry food is limited. The oil content accounts for its exceptionally high metabolizable energy value.

Sunflower oil meal

The decorticated meal has a fibre content of approximately 14% and the undercorticated product about 30% fibre. Sunflower seed oil meal is very palatable and keeps well in storage. When it has been over-heated it is very deficient in lysine, due to the destruction of this amino acid.

Linseed meal from flax (*Linum usitatissimum* L.)

The protein of linseed meal does not effectively supplement the protein of maize or other grains for poultry. It is not satisfactory when forming an appreciable part of the rations for poultry. Linseed meal has a fair calcium content, averaging 0.37% and is rich in phosphorous with an average of 0.86%.

Animal protein foods

Animal protein foods are those animal and fish products that are either surplus to or unfit for human consumption. Care should be taken with their use since they may be a source of disease. In developing countries they are frequently either unavailable or very expensive. Examples of these foods available in tropical countries are as follows.

Fish-meal

Fish-meal is produced by drying and grinding fish from either salt water or fresh water sources. Fish-meal is a high quality protein food rich in all the essential amino acids, vitamin B_{12} and choline. It is a good source of riboflavin, calcium and phosphorus. Fish-meal is in limited supply in most developing countries. It may be possible to produce supplies locally. When stored in hot climates it can deteriorate in quality and should therefore be protected by the addition of an antioxidant. High inclusion rates (above 10%) may lead to taints in poultry meat and eggs, especially if low quality fish-meal is used.

Meat and bone-meal

This is usually made from the carcasses of animals that die on the farm. The whole animal, except the skin and hooves, is cut up, cooked, dried, and then ground to form a meal. The quality of the meal will depend on the relative proportions of meat and bone in the sample. Meat and bone-meal that is to be used in poultry diets should preferably contain at least 50% crude protein. It is a good source of lysine, methionine, cysteine and tryptophan and it is also rich in calcium and phosphorus. Because of its high oil content it can easily deteriorate in hot climates, especially hot humid ones such as those in the states bordering the Gulf of Guinea, and it should therefore be protected by the addition of an antioxidant.

Feather meal

Producers of large numbers of broilers can usefully dispense with the feathers produced by preparing feather meal. This is best produced by processing the feathers in steam and grinding the product. Feather meal is a rich source of protein high in leucine and cysteine but deficient in lysine, tryptophan and methionine.

> *Beware of the dangers of infecting birds with disease, particularly those caused by salmonella bacteria.*

Other animal proteins

Other animal proteins that could be used for feeding to poultry include blood-meal and poultry dressing plant by-product meal.

> *With these products there are dangers of infecting both birds and eggs with disease unless they are thoroughly sterilised before incorporating into food.*

5 Intensive systems of egg production

Introduction

Poultry of various classes are kept for egg production, these include the hen (*Gallus domesticus*), the duck, the guinea fowl and the goose. Of these, only the first two are of major importance and the vast majority of eggs produced for human consumption are produced by hens. Therefore most of this chapter will be concerned with this species although the last section will be concerned with egg production from turkeys and from ducks. The system dealt with will concern intensive management; the rearing of chicks under field conditions will be dealt with under the section concerned with scavenging poultry.

Stress and welfare

In many western countries a considerable percentage of the population is becoming concerned about the concept of factory farming. This means the keeping of birds under intensive conditions particularly the keeping of birds in cages. Such forms of production are regarded by some people as inhuman and stressful. In Britain as a result of the Brambell report in the 1960's various codes of practice were introduced. These outline the right of birds to water and food, adequate ventilation and temperature, space, light, diagnosis and treatment of injuries and disease. In Switzerland the use of battery cages for laying hens is now prohibited. However such forms of management are undoubtedly productive and profitable from the farmer's point of view so even if the birds are stressed their productivity would not seem to be unduly affected.

The effects of the environment on the welfare of animals can be divided into physical and social effects. Direct physical effects may result in the ill-health of animals and thus poor welfare. The social effects are less easy to describe. All domestic animals are gregarious, living in groups with an organised social structure. The constraining of animals and birds in intensive systems is bound to affect their social organisation.

Stress is an inevitable aspect of the lives of all animals and man, and it is unreasonable to expect the lives of birds to be free of stress. The question we must ask ourselves is to what extent should birds be exploited. Some forms of suffering, such as feather pecking and cannibalism in birds are obvious to the observer. The effects of the social environment on birds are much more difficult to determine. Careful experiments are needed to obtain knowledge about the feelings of birds. All forms of poultry production modify the bonds between individual birds. By taking account of the social interactions between birds the deleterious effects of intensive systems on behaviour can be minimised.

Behavioural and physical evaluators of stress and distress are still inadequate to identify 'ideal' production environments. Extreme stress results in lowered productivity of individual birds. The maintenance processes of a bird have a higher priority than productive processes, so that production is a sensitive measure of stress. A well-designed system and good stockmanship are the keys to both low stress and high production.

Rearing chicks to eight weeks of age

Chickens are normally reared from one day old in brooder houses which should be sited well away (at least 100 m), and preferably on another farm, from those which house adult birds. Disease is very easily passed from adults to young chickens and this problem is best avoided by housing only birds of the same age on any one farm (the all-in/all-out principle) and by ensuring that operatives attend only to their own birds and do not carry disease from one group of birds to another. The accommodation should be rested between batches (2–3 weeks) and adequately cleaned and disinfected to reduce the risk of carry-over of disease from one batch to the next.

The best method is to rear the chicks on deep litter (although alternative methods such as tier brooders are available). Where deep litter rearing is practised, 6–12 cm of clean litter should be spread on the floor and brooders turned on twenty-four hours before the birds arrive to ensure that the brooder house and the brooders are at the correct temperature (Table 5.1).

Table 5.1 Temperature requirement of chicks

Age of chicks	Temperature under hover (°C)	Temperature of building (°C)
Day-old – 1 week of age	35	
1–2 weeks	30	(minimum 15)
2–3 weeks	26	
3–4 weeks	23	

Fig 5.1 *Simple paraffin fuelled hovers being used to rear turkey poults (note that the poults are eating the food from keyes trays)*

The heating is usually provided from an overhead source such as paraffin heaters (Fig 5.1), infra-red lamps, gas heaters or steam pipes. Whichever system is used it is important that a back-up system be provided in case of power failure. Young chickens are extremely vulnerable to cold.

Food and water should be put out before the chicks arrive. To encourage the chicks to eat, the food should be put into keyes trays or shallow baskets for the first day or so and after this into the normal food troughs.

The chicks should be confined to the area of artificial heat for one week otherwise they may stray and become cold and chilled. Each bird should be able to reach a heated spot under the brooder and it is necessary therefore to provide 40–60 cm^2 of brooder space per chick. Behaviour of the chick is a good indication of whether the temperature under the brooder is too low or too high. If all the chicks are crowded under the brooder the temperature is too low (Fig 5.2). If they are crowded around the cardboard surround the temperature is too high. If they move to one side of the pen there is probably a draught. If the birds are evenly spread and appear content, the temperature is probably about right.

It is also important that the birds are not overcrowded, overcrowding can lead to slow growth rates, disease build-up, feather pecking and

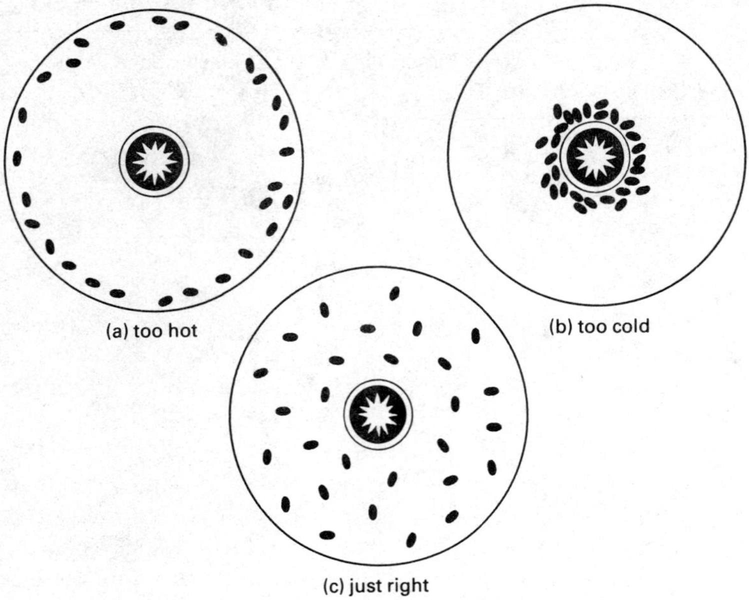

Fig 5.2 *Diagram of the behaviour of chicks when a brooder is (a) too hot, (b) too cold and (c) just right* (Chris McKinnell)

cannibalism. Feather pecking can be minimised by trimming the chicks' beaks (Fig 5.3). As the chicks grow so their space requirement grows (Table 5.2).

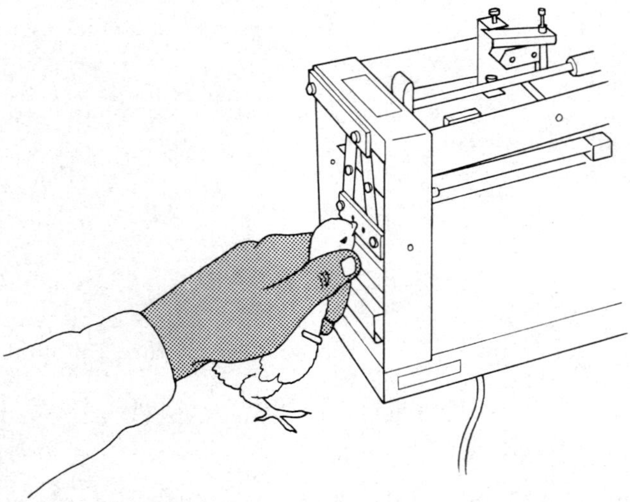

Fig 5.3 *Trimming the beak of a chick* (courtesey of *ATB*)

Table 5.2 Floor space and trough space requirement per 100 chicks

Age (weeks)	Floor space (m^2)	Trough space (m)
0–4	4	1.5
5–8	9	3.0
9–20	12	6.0

The food fed initially should be in the form of crumbs. In addition to the food a five litre water fount should be provided for every 100 chickens. Care should be taken to ensure that the litter around the fountain remains dry.

The birds should be treated with Marek's vaccine when they are still under one day of age. At four to six days of age they should be vaccinated against infectious bronchitis and Newcastle disease. The infectious bronchitis treatment should be repeated when the birds are three to four weeks old. There is not a vaccine against coccidiosis therefore the chick food should contain a coccidiostat.

Rearing potential laying birds from eight weeks of age

This is a fairly straightforward operation if the poultry are housed under intensive conditions. The poultry are normally kept in the same house that was used for the initial rearing period but the operatives have to be sure that enough floor space and feeding space is available (Table 5.2).

At six to eight weeks of age the birds are changed from a chick diet to one suitable for older birds. This diet is lower in protein and consequently cheaper. Probably the best way to effect the change is to allow a fixed amount of chick diet to be provided for the first six weeks. Once all this food has been eaten the birds will be gradually changed onto a grower's diet. This diet need not contain a coccidiostat. Wastage of food during the growing phase can be a very serious problem. It is very important that food troughs should not be overfilled neither should the tube feeders be too widely opened (Figs 5.4 and 5.5). The construction of the food trough is also important and there exist types that reduce spillage and so reduce food wastage by up to 20%. Food wastage can turn a profitable enterprise into one making a considerable loss.

In areas where fowl pox is a problem the birds should be vaccinated against this disease when they are four to twelve weeks old. An additional vaccination against Newcastle disease should take place at sixteen weeks of age.

Fig 5.4 *Growing birds in pens where the food is provided in tube feeders*

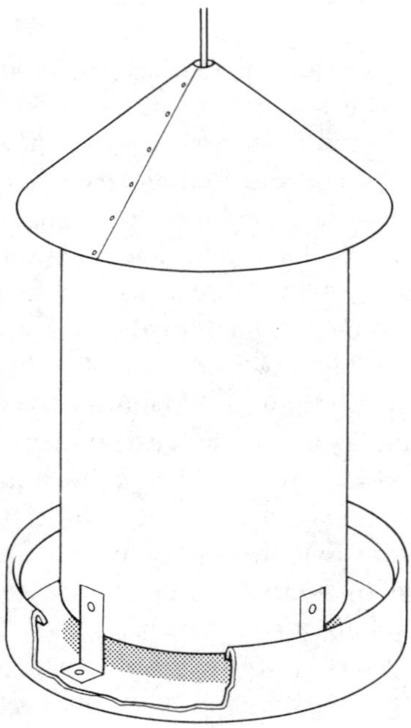

Fig 5.5 *Tube feeder — with portion cut away to show detail* (Chris McKinnell)

A careful watch should be kept for signs of feather pecking. This can often start in the late afternoon when the sun shines on the feathers of the birds. Badly pecked birds should be treated with Stockholm tar.

Day-length should be reduced from twenty-four hours to fourteen-and-a-half hours or less to delay maturity and the onset of egg production.

The laying phase

Egg production mechanism

Only the left ovary is functional in laying hens (if the right one becomes functional it normally produces testosterone causing the birds to cease production and grow large bright red combs).

The immature ovary contains many hundreds of minute yolks and at sexual maturity some of these yolks start to develop. The ovary of the laying hen will contain several yolks in various stages of development. Each yolk is enclosed in a thin membrane called the vitelline membrane. Around this membrane is a sac or follicle which supplies the yolk with nourishment by means of a network of tiny blood vessels. The yolks mature seven to ten days after they start to grow and at this stage the

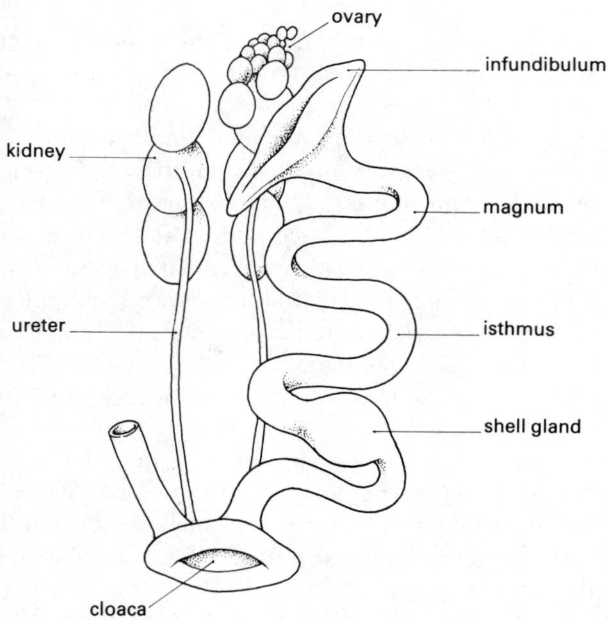

Fig 5.6 *Uro-genital system of a laying fowl – the reproductive system has been removed from the left side of the diagram to show detail of the kidney*

yolk slips out of the follicle and into the funnel shaped opening of the oviduct (the infundibulum). It remains in the infundibulum (Fig 5.6) for approximately eighteen minutes and if the hen has been previously mated, it is at this stage that the ovum is fertilised.

From the infundibulum the egg descends to the magnum region where the layers of albumen are secreted around the yolk. It remains in this region for approximately three hours. It next moves to the isthmus region of the oviduct where the two shell membranes are secreted around the albumen. It remains in this region for one-and-a-quarter hours. Lastly the egg descends into the oval shaped uterus region where it remains for twenty to twenty-one hours. In the uterus more albumen and water are added so the egg is plumped to its conventional egg shape. In addition the shell is laid down around the egg in the uterus. This is a very slow process taking approximately eighteen hours. This is one of the main constraints to more rapid egg production. Sometimes, particularly with young pullets, two eggs are shed into the uterus at the same time resulting in the formation of a double yolked egg. One shell being secreted around two eggs at the same time. Alternatively only one egg will be surrounded by a shell and one egg will be laid without a shell, a soft-shelled egg.

Management

Birds of egg laying strains are normally moved to their laying quarters at eighteen weeks of age. These quarters may be on a separate farm or should be at least 100 metres from the houses where the birds are reared. Many types of intensive housing are used for laying birds and these include battery cages, slatted floor housing, deep litter housing and more recently the aviary type house. All these types have advantages and disadvantages which will be discussed later. The following management points apply to all types of intensive management. Types of semi-intensive housing are shown in Figs 5.7, 5.8 and 5.9. Space requirements of birds under these systems are shown in Table 5.3.

The dimensions of a simple battery cage are given in Fig 5.10.

Chickens can also be kept in countries with a high rainfall on pebble yards (Fig 5.11). The rain washes the pebbles clean.

A more common system of semi-intensive production is the half-inside/ half-outside system (Fig 5.12). This system has few of the advantages of either intensive or of scavenger production although the housing if correctly designed may reduce heat stress on the birds in the hot season of the year.

A field ark which can be moved onto clean pastures each day (Fig 5.13) has advantages when land is plentiful and predators are not a problem.

Table 5.3 Space requirements of laying birds kept under different systems of management

System	Floor space (m^2)	Perch space (cm)	Nest space	Feeding space (cm)	Drinking space (cm)	One man unit
Semi-intensive	0.15 solid floor 0.10 slatted floor	20 heavy breed 16 light breed	trap nests: 3 birds/nest open nests: 5 birds/nest	12	5	1000
Straw yard	0.20 house 0.40 yard	20 heavy breed 16 light breed	communal nests 2.0 m^2/100 birds	12	5	2500
Deep litter	0.25 light breed 0.30 heavy breed	20 heavy breed 16 light breed	communal nests 2.0 m^2/100 birds	12	5	4000
Battery	0.10	nil	nil	12	5	4000
Breeding ducks	0.60	nil	2.0 m^2/100 birds	12	5	2500

Fig 5.7 *Battery house (Californian type) (left)*

Fig 5.8 *House with slatted floor (right)*

Fig 5.9 *Intensive battery house*

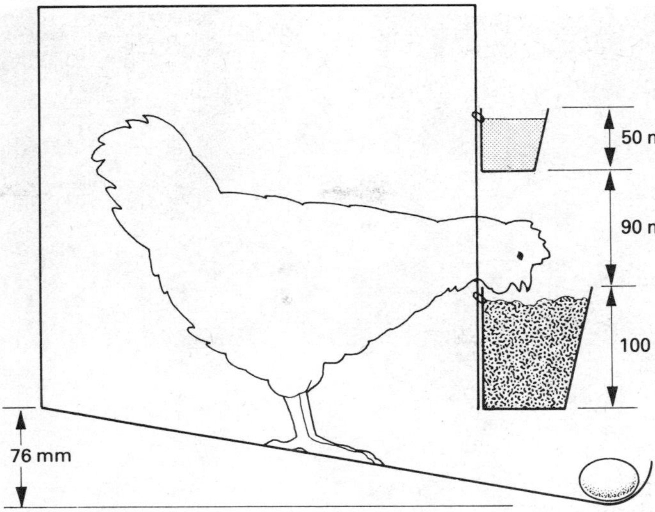

Fig 5.10 *Dimensions of a simple battery cage* (Chris McKinnell)

Fig 5.11 *Chickens on a 'pebble yard' – a self-cleaning system in high rainfall areas (left)*

Fig 5.12 *Semi-intensive system of poultry keeping for egg production in Zimbabwe (right)*

Fig 5.13 *Field unit (movable fold) for a laying bird*

Feeding

At the time that the birds are moved to their laying quarters their diet should be changed to one that contains a high level of calcium (3–4%) or alternatively supplemental calcium should be made available in the form of calcium grit (for example oyster shell grit). The banks of the river Niger have accumulations of the shells of mussels that are used by local farmers as a source of calcium for poultry. A high level of calcium in the diet should be fed for two weeks prior to the commencement of lay because at this time the change in the hormonal status of the birds enables them to deposit extra calcium in their bones (especially in the medullary bones [1]). This extra calcium is mobilised when the birds start to lay eggs. It is normal for birds to be fed *ad libitum* at this time since any form of rationing will result in reduced rates of egg production.

Water

Water supplies to laying birds in the tropics are crucial and any lack of water will result in decreased egg production and possibly death.

Lighting

In order to stimulate egg production the length of the artificial day used should be increased either in one step, or in a number of steps until the artificial day reaches sixteen to eighteen hours (Fig 5.14). This will

1 Medullary bone is a special bone tissue formation located within the medulla of many bones in laying chickens. This bone structure is easily mobilised for use in shell formation.

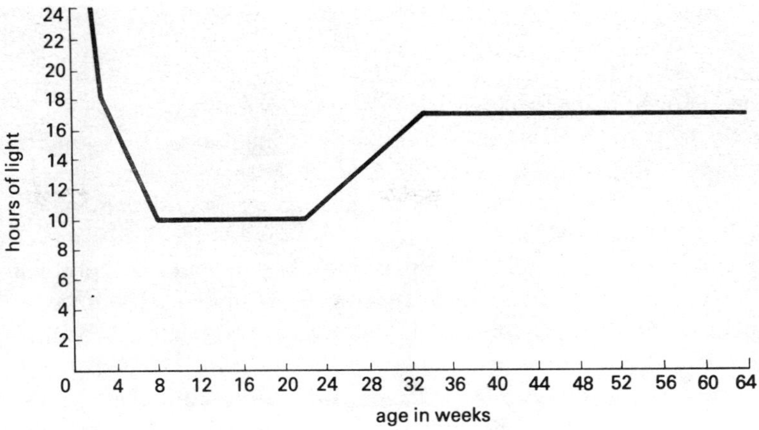

Fig. 5.14 *Ideal lighting curve for hens kept under 'environmentally' controlled conditions* (courtesy of *ATB*)

ensure that maximum production is achieved in the shortest possible time and that egg production is maintained at the highest possible rate. An ideal lighting curve for hens kept in windowless houses is shown in Fig 5.14. In the type of open housing that is commonly used in the tropics, the provision of one hour of artificial light at 03.30 h and at 19.30 h has been shown to result in levels of egg production not very different from a conventional programme in which lighting (natural or artificial) is provided from 03.30–20.30 h. The former system would result in lower electricity bills.

Under these conditions of lighting a bird should produce 250 to 280 eggs in a manner illustrated by the normal laying curve (Fig 5.15).

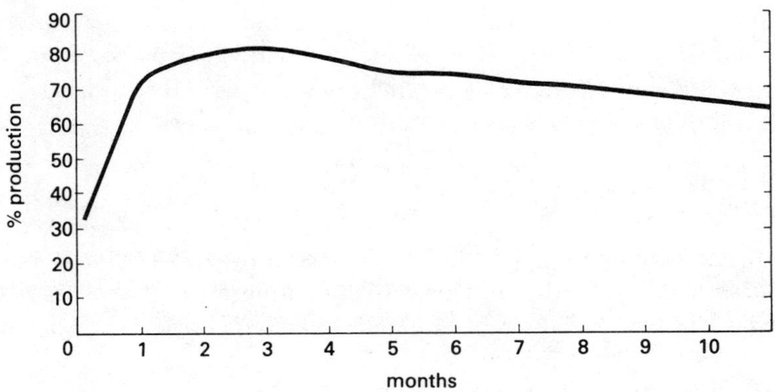

Fig 5.15 *Typical egg production curve for high producing hybrid poultry* (courtesy of *ATB*)

Broodiness

Under natural conditions hens will lay a clutch of say twelve eggs and they then go broody[1] so that these eggs can be incubated. This condition is associated with increased production of the hormone prolactin and it is prevented by the secretion of oestrogen. Under intensive management broodiness is obviously a disadvantage since broody birds do not produce eggs and therefore considerable steps are taken to avoid this problem. It is most likely to occur in birds kept on deep litter and can be a real problem when the eggs are laid in communal nest boxes. Broodiness is encouraged by these boxes and if broody hens are allowed to remain they may break large numbers of eggs when trying to make a nest. This can lead to an outbreak of egg eating by other non-broody birds. Thus it is desirable that broody birds be removed from the colony as soon as possible. If this is done and the birds are put into wire-floored pens with free ventilation they will lose their broodiness and return to production. This process will be aided by adequate feeding and by long day-length.

Luckily broodiness is an inherited characteristic and breeders have bred the trait out of modern hybrid layers. In non-hybrid stocks heavy breeds tend to go broody more frequently than light breeds.

Moulting

Laying birds normally moult at the end of the laying year and in this process lose and regain a new set of feathers. The process is initiated by an increased production of the thyroid hormone, thyroxine. It can be prevented experimentally by the injection of anterior pituitary extract. Uncontrolled moulting is an undesirable process which is associated with the cessation of egg production. It is initiated by a decrease in day-length or by the stress caused by a lack of water and/or food. In the hot season of the year the lack of water is a cause of accidental moulting by birds and a complete cessation of egg production. Birds that are to be kept for a second layer year are sometimes force moulted to ensure that they all resume laying at the same time.

Culling laying birds

Battery birds should be culled[2] all the time, deep litter birds as often as is convenient. Culling from batteries is easy, since the performance of individual birds can be recorded. In deep litter use the following points to decide which birds should be culled.

- Cull birds that are sick.

1 Broodiness is a natural mechanism found in all birds which causes them to want to sit on and incubate eggs until they hatch.

2 Culling means removing unproductive birds from the flock.

- Cull birds that do not feed.
- Cull birds that are too fat or too thin.
- Bluish coloured combs indicate ill health.
- Good birds will have scruffy feathers, unproductive birds may have a well preserved appearance.
- Pigment in yellow fleshed birds is lost first from the vent, eyelid, ear, beak and lastly legs. This loss of pigmentation can give an indication of productivity. The lower the pigmentation the greater the productivity.
- Birds with dry vents are normally not producing.
- Birds should have room for two or three fingers between pubic bones and for four between pubic bone and breast bone (Fig 5.16).

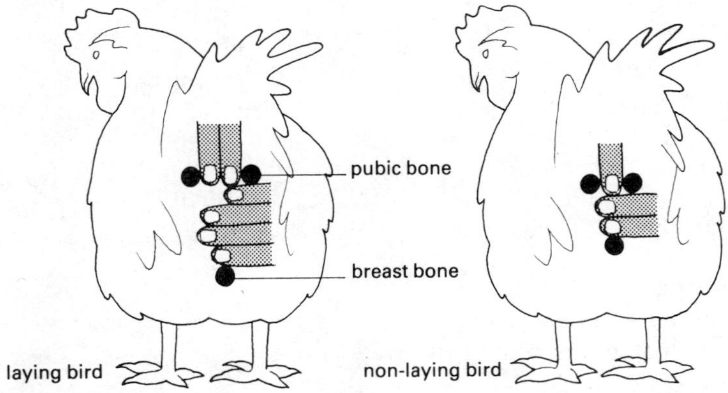

Fig 5.16 *A simple means of distinguishing laying birds from non-laying birds* (courtesy of *ATB*)

Egg quality

When eggs are being produced for human consumption it is important that they are suitable for this purpose. This will be determined by both the external and internal qualities of the eggs.

External qualities

Shape
The eggs should be of the conventional egg shape. Round or long eggs are undesirable as they are not liked by commercial poultry keepers because they are difficult to pack into keyes trays and are liable to become broken. However in tropical countries eggs with deformed shells are readily sold at a lower price than normal eggs.

Shell quality

The shells of eggs must stand up to handling. A broken egg cannot be sold. There is a strong correlation between shell strength and shell thickness. Eggs with a shell thickness of 25 micrometres or less are invariably broken in the handling process. The ideal shell thickness is about 35 micrometres. Shell thickness can easily be determined by the use of specific gravity.

The mean thickness of shells is determined by many factors including the individual physiology of the bird, air temperature, nutrition and disease. Infectious bronchitis has a particular effect on shell development.

Shell colour

The colour of the egg shell is determined by a thin layer of oöporphyrin. The colour of shells has no effect on the nutritional quality of the egg but in several countries, notably in the UK, consumers have preferred brown-shelled eggs to white-shelled ones and have paid a premium for them. Even though breeds producing brown-shelled eggs are less productive than those that produce white-shelled ones, breeders over the last ten years have made a concerted effort to produce highly productive strains of brown egg producers. This task has been made doubly difficult by the fact that the more eggs a bird produces the lighter the colour of the egg shell.

Internal characteristics

There is nothing more likely to put a consumer off eggs than for them to be given eggs with substandard internal quality. It is therefore most important that a high quality should be maintained if sales are to be maintained. The internal characteristics that should be considered are as follows.

Blood and meat spots

Blood spots are caused by blood being released into the yolk when the follicle ruptures from the ovary. Meat spots normally come from the oviduct and are most commonly found in the albumen (Fig 5.17). Neither is harmful to the consumer but their presence is not likely to impress him or her. The predisposition to both characters is inherited and the amount of these faults can be reduced by selection of stock. Blood spots are more common in birds fed on high levels of vitamin K.

Yolk colour

The colour of the yolk is determined by the presence or absence of xanthophylls some of which are not precursors of vitamin A. The colour of the yolk is not therefore a guide to the yolk's nutritional value, since

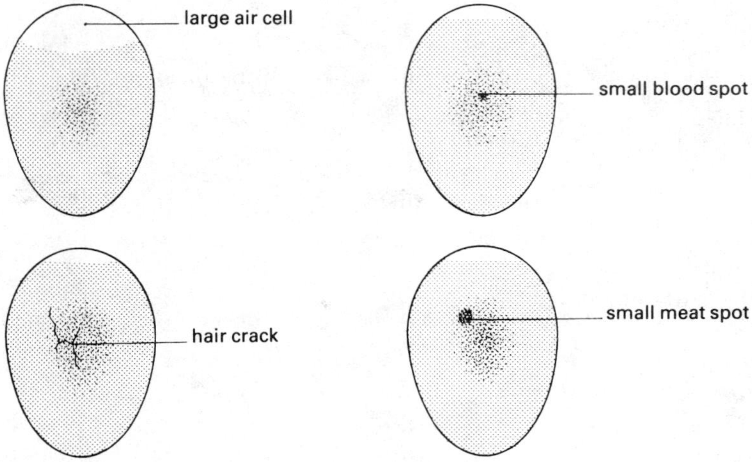

Fig 5.17 *Egg faults*

vitamin A itself is colourless. The colour of the yolk is influenced to a large degree by nutrition and dark yellow yolks can be produced by feeding laying birds on grass meal or alfalfa meal. In fact it is possible to influence the colour of the yolk to a very precise degree by the feeding of artificial pigments such as canthaxanthine and β apocarotenoid. The colour of the yolk can be specified by comparison with the Roche fan.

Proportion of thick white

Eggs containing a large proportion of thick white are normally regarded as being of high quality. This is because on storage the amount of thick white decreases and thin white increases. Therefore there is a natural association between thick white and quality (Fig 5.18). However the proportion of thick white is also influenced by other factors. For example there are differences between the eggs laid by individual birds. Certain diseases, notably infectious bronchitis, reduce the proportion of thick white.

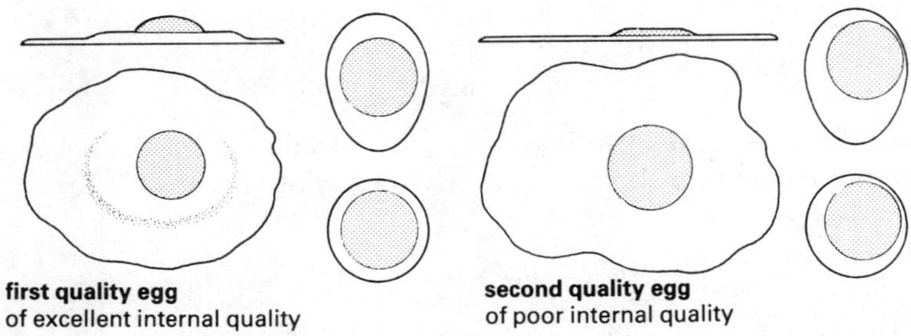

Fig 5.18 *Egg quality*

Taints

Stale eggs will have undesirable taints due to rotting. Eggs which have been stored for a long time are more likely to be unsound. This can be determined by immersing the eggs in water. Those that float are more likely to be old and unsound.

Tainting may also be caused by feeding second quality fish-meal. However, eggs are rather like milk in that they will absorb smells from the atmosphere. Thus if eggs are stored in a smelly atmosphere, whether the smell is from a newly painted room or from fish-meal, they are likely to become tainted.

Clean egg production

In order that eggs should reach the consumer in a fresh and palatable state it is important that the eggs be produced under as clean conditions as possible. Dirty eggs even if subsequently cleaned, are likely to keep less well than eggs produced under clean conditions. Eggs can be kept clean by keeping the equipment as clean as possible and by collecting the eggs as frequently as possible (at least three times a day). Frequent collection avoids eggs being broken and the contents of broken eggs coating the unbroken ones.

Cleaning eggs

There are three methods of cleaning eggs:

- dry cleaning;
- cleaning with a damp rag;
- immersion cleaning.

The first (dry cleaning) is impractical because with a mechanical system cracked eggs or thin shelled eggs break and make following eggs dirty. The second (damp rag) spreads disease from one egg to the next. The third is the most satisfactory, particularly if the water used for cleaning is warm and contains a detergent and sanitiser. The water should always be warmer than the eggs. Before cleaning, eggs should be sorted into clean and dirty and only the dirty ones should be washed. This is best done in an egg washing machine of which there are several types on the market.

Storing eggs in water glass

To store 100 eggs a 25 litre jar is required. Mix 6 litres of previously boiled cool water with 0.5 litres of water glass (sodium silicate). Place the eggs in the jar and cover them with the water glass solution, cover the jar and place in a cool shaded spot. The eggs should stay edible for up to six months. Check the effectiveness of the storage by removing two eggs every month and assess them for freshness.

Broiler breeders

Broiler breeders are laying hens that have been bred especially for the production of chicks that will be fattened as broilers. Selection has taken place for rapid growth rate and efficient food conversion and therefore these hens are much heavier than the birds used for the production of eggs for human consumption. Also because egg production and growth in body size tend to be negatively correlated they also produce fewer eggs (140 per annum compared with 250). The management of these birds is similar to that of egg laying birds except that because the birds are used for breeding, cockerels must be kept with the hens in the laying stages and therefore the deep litter system of management is most convenient.

Rearing broiler breeders (parent stock)

A similar system is used as in the rearing of laying bird chicks. The same vaccination procedure is followed but in addition avian encephalomyelitis vaccine should be given in the water when the chicks are eighteen weeks of age. Chicks of poor quality should be culled at six weeks of age; 3–5% of the hatch will normally be culled at this stage. The lighting pattern should be similar to laying birds.

The main problem with immature broiler breeders is that they tend to eat too much, grow too quickly and become obese and therefore various techniques have been devised to prevent this happening, these techniques include restricting the daily amount of food fed, low protein diet, high fibre diet, skip-a-day feeding, low lysine diets and restricting time of access to food. These various techniques are discussed in more detail in Chapter 4.

It is probably best to rear the cockerels and the hens separately for several reasons including the fact that cockerels have a different nutrient requirement to that of hens. The cockerels need more calcium, phosphorous and some other nutrients initially. The two sexes should be mixed at twelve to fourteen weeks of age at a ratio of 8–10 hens to every cockerel.

All pullets and cockerels should be removed to the laying quarters at the same time. The quarters should be situated well away from the area where the immature birds are kept. This latter area should be thoroughly cleaned and should not receive new chicks for at least three weeks after the previous flock of hens have been removed. Outside runs which are very difficult to clean should be left vacant for at least two months between batches of birds.

Management of broiler breeders in lay

Both pullets and cockerels should be moved to the laying house at twenty-one weeks of age. The lighting should be as for laying birds. The birds should reach peak production at thirty to thirty-three weeks of age.

Care must be taken with respect to heat stress in the tropics because broiler breeders are more susceptible to heat stress than birds of laying strains.

Breeding turkeys

The management of breeding turkeys is influenced by the type of stock used. Traditional stock can be allowed to mate naturally but modern hybrid stocks have to be bred by artificial insemination (AI). The hens need to be kept under a short day-length regime in order to initiate egg production. The egg production by female turkeys is less than that by broiler breeders, i.e. some 70–130 eggs per year.

AI is used because modern strains of turkeys are so cumbersome and broad-breasted that natural mating is difficult if not impossible. The males have extreme breast widths and are not able to perform natural mating, the stag losing its balance before inseminating the semen. Under these conditions the use of AI is essential. Turkey semen cannot be stored for more than thirty minutes and therefore the stags have to be 'milked' immediately before inseminating the hens. The hens are inseminated once every one to two weeks with 0.05 cm^3 per hen.

In developing countries where skilled labour may be unavailable it may pay to use less extreme types of turkey that can mate naturally. With a natural mating system the ratio of hens to stags should be about 10:1.

Breeding turkeys are sensitive to changes in day-light and can be induced to produce by artificial changes in day-length. Hens can be prevented from laying by maintaining them on a short day-length pattern (eight hours) during the rearing period. Hens are more sensitive to day-length changes than are stags and therefore stags should be exposed to an increasing light regime two weeks before the hens.

Producing eggs from ducks

The advantages and disadvantages of ducks

Ducks, like chickens, are kept for two distinct purposes, meat production and egg production. The distinction may be less marked in developing countries.

The birds used for these two purposes come from different sources. The Aylesbury and Pekin and strains derived from them are used for meat production whilst the Khaki Campbell and strains derived from it are used for egg production.

Ducks used for meat production have a very fast growth rate – modern hybrids will achieve 3.3 kg liveweight at fifty-three days of age (compared with 2 kg for a broiler of a similar age). They have a high reproductive rate, similar to that of hybrid chickens even though they have been subjected to less selection pressure for this trait than have chickens. They are very hardy and are much less susceptible to disease than chickens. They can be removed from the brooder at two weeks of age even in temperate zones and because they suffer from few diseases they can be reared on a 'conveyor system'. In this system the ducks start at one end of the house at the time of hatching, are moved on a pen a week later and are removed from the other end seven to eight weeks later.

A disadvantage is that they have a high fat content in their carcass, particularly subcutaneous fat, and as a consequence tend to use food inefficiently. This inefficiency is made worse because they tend to be very wasteful feeders, pushing a considerable quantity of food out of the trough.

Ducks are also very nervous and have a tendency to panic. They move around in flocks and if they are frightened when they are in a closed building, large numbers of them can be trampled underfoot in the ensuing panic. Because of this problem, in the past ducks have been kept outside, even in temperate regions. However it is now known that provided care is taken, there is no reason why ducks should not be kept inside even when kept in large numbers. Water is not required for swimming or mating and provided ducks can immerse their heads in water from time to time they will remain perfectly healthy.

Egg production from meat producing ducks

The breeds that are used for producing table ducklings are the Aylesbury and the Pekin. Females of both breeds are capable of producing 180 eggs per bird per year. The Aylesbury is a white plumaged, white fleshed bird, the drakes of this breed weigh about 4.5 kg and the duck 4.0 kg.

Table 5.4 Relative productivity of hens and ducks kept for egg production

	Brown egg laying hen	White egg laying hen	Khaki Campbell	Cherry Valley laying duck
Hen housed average (eggs/bird/year)	269	277	237	285
Mean egg weight (g)	59.70	56.90	75.50	75.00
Food consumed:				
0–20 weeks (kg)	8.00	7.10	8.90	11.20
21–74 weeks (kg)	42.20	41.50	56.80	43.70
Total food/kg egg	3.13	3.10	3.67	2.56
Point of lay weight (kg)	1.61	1.30	1.45	1.50

After J. Powell (Cherry Valley)

The Pekin has a creamy-white plumage and yellow flesh and both sexes are slightly lighter than the Aylesbury (4.0 kg and 3.6 kg respectively).

Breeding stock should be housed together at least one month before their eggs are required for selling. Normally first year ducks are used to produce the eggs. One drake is kept with approximately five ducks, but flock-mating is a more normal practice with six to seven drakes kept with twenty-five to thirty ducks.

Ducks are light sensitive and are induced to commence production by an increasing light regime. Long artificial days are provided at twenty weeks of age and the birds should commence laying two weeks later. Drakes should be stimulated by artificial light two to three weeks before the ducks to ensure that they produce fertile eggs. It is difficult, even with artificial light to maintain egg production by egg/meat type ducks and therefore a number of flocks should commence laying at different times of the year to ensure continuous production. The stocking rate should not exceed two per square metre. Although these breeds can lay more eggs, not more than sixty to seventy live ducklings should be anticipated from each female duck in the flock.

Egg production from egg producing strains of ducks

Management of egg producing ducks

Ducks of egg laying strains can produce as many eggs as productive hybrid poultry giving flock averages in excess of 270 eggs per year (Table 5.4). They start laying at four and a half months of age and lay

Table 5.5 The nutrient content of eggs produced by chicken and ducks

	Chickens	Ducks
Protein content %	12.9	13.5
Fat %	10.9	14.5
Ash %	0.9	1.0
Dry matter %	26.4	30.3
Average weight (g)	57.0	75.0

After J. Powell (Cherry Valley)

for twelve months or more. They can be kept in cages but on the wire floors of cages their feet may become damaged. They can be kept in deep litter houses in temperate zones. In tropical countries the best results are obtained if the birds are kept outside, with or without a pond for swimming. If a pond is not provided, basins in which they can immerse their heads should be available. The eggs produced weigh on average 60 g and are almost white in colour. The eggs have a slightly higher nutritive value than hens' eggs (Table 5.5). Ducks are immune to most disease problems prevalent amongst gallinaceous birds and no vaccinations are needed. They are susceptible to duck hepatitis, cholera and liver fluke and are particularly vulnerable to flavotoxins produced in groundnut by *Aspergillus flavus*.

Rearing is simple, less heat and a lower protein diet are needed than for chicks. Forty square metres of roofed area should be provided for each one hundred laying ducks with a run of double this area. Nests should be provided because, given a choice, ducks will always lay in a dark secluded place with a soft floor which should of course be dry and clean. There should be water deep enough to allow the birds to immerse their heads.

It is not desirable to bring ducks into lay too soon or else they will produce too many small eggs. The flocks should be laying at 50% four to five weeks after the first egg is collected and they will reach a peak of production, over 90%, within 8–10 weeks. High producing ducks have large clutch sizes. The clutch size is a heritable trait.

Broodiness, another cause of lack of production, is heritable as is early moult. Ducks are very nervous and sensitive to any change in routine, food, feeding time and light schedule. Any of these can induce early moulting. Ducks lay their eggs early in the morning, 97% by three hours after the artificial lights are turned on.

6 Intensive systems of poultry meat production

Introduction

The broiler industry is the best integrated one in the whole of the field of animal agriculture. The term **broiler** is an American one and is applied to chicks of the species *Gallus domesticus* that have been bred especially for rapid growth so that they will on average reach a weight in excess of 2 kg at eight weeks of age having consumed only 2 kg of food for each kilogram of liveweight increase (Fig 6.1).

The strains of birds are very different from the types used for egg production which are normally based on the White Leghorn; broiler strains are based on crosses between Cornish White, New Hampshire and White Plymouth Rock.

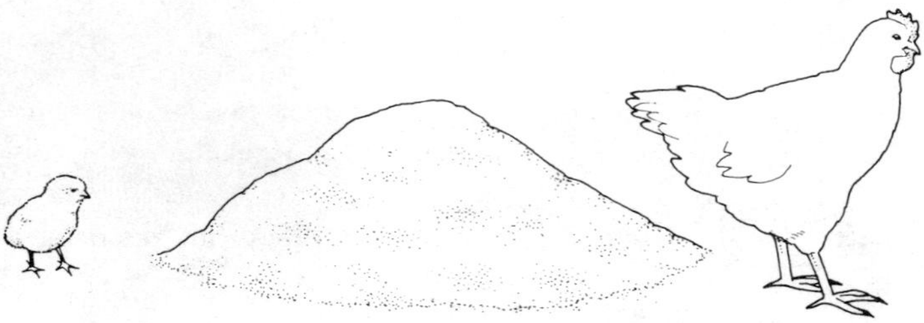

Fig 6.1 *A one week old chick and a seven week old broiler, with the amount of food required to turn the chick into the broiler*

Broiler production

The main guiding principle of broiler rearing is the 'all-in/all-out' principle so that only birds of the same age are kept on the same site. The birds can be reared to slaughter weight in eight weeks, and two weeks are required between each batch so five batches can be reared in each house each year. On modern farms, birds are kept in houses holding between 10 000 and 20 000 birds (Fig 6.2) and one man should be able to manage five houses, provided help is available for catching and crating the finished birds and for cleaning out and refitting the houses for subsequent batches.

Fig 6.2 *An intensive broiler house at Belo Horizonte in Brazil*

The broilers are reared in a similar way to immature layers except that the system is almost always on litter. Special care has to be taken with the litter, since caked litter can lead to the formation of breast blisters and the down-grading of carcasses. The birds are kept at very high densities, sometimes as high as twenty birds to the square metre. In temperate zones when the birds are kept in windowless houses with fan ventilation the main problems are keeping the birds warm enough when they are young and removing the excess heat when they are older. Emergency power supplies are a must because power failure can be a disaster. In certain underdeveloped countries, failure of the power supply can be frequent and long in duration, and so a self-contained generator is essential if this type of house is used. However where open-sided houses are used such services are not essential but even so deaths from heat stress can be a problem.

Broilers in developing countries are normally run 'as hatched'. In developed countries because the cockerel and hen have different nutrient requirements they are often put into different pens. There are advantages in running the two sexes together because broilers run as hatched will produce a variety of birds at eight weeks of age which will suit the various demands of the consumer. At eight weeks of age the males will usually weigh 200 g or so heavier than the females and will have had a better food conversion ratio. Females on the other hand produce a more attractive carcass which may fetch a higher price.

In both sexes the best food conversion occurs at the earliest stages of growth and declines thereafter. Thus the cost per unit of gain increases steeply with age despite the fact that the birds can be fed on lower protein rations (and therefore cheaper rations) once they reach four weeks of age. The reason for the decline in efficiency is twofold, firstly as the birds get larger so they have a higher maintenance requirement and secondly because as they mature so the proportion of fat laid in the carcass increases. It requires far more energy to produce a kilogram of fat than it does to produce a kilogram of lean.

Disease control

A similar control programme is advocated as for laying chickens. The birds are vaccinated against Marek's disease at one day old, for Newcastle and infectious bronchitis at four to six days and twenty-one to twenty-eight days respectively. A second Newcastle vaccination is given at forty days of age. The food should contain a coccidiostat which should be removed four days before the birds are killed.

Feeding and water supply

The objective is rapid growth rate and therefore it is most important that adequate trough space is provided. Two-and-a-half metres per 100 chicks from nought to three weeks and six metres per 100 chicks from four to eight weeks. Maximum intake is assured if the birds are fed on pellets rather than mash (Fig 6.3). If the food troughs are present in insufficient numbers or are badly maintained, 20% or more of the food may be wasted. Water must also be provided at all times (see Chapter 5).

Fig 6.3 *Automatic feeding system for broilers in use in Brazil*

Lighting

When broilers are kept in windowless houses both day-length and light intensity can be adequately controlled. The same degree of control is not possible in the open-sided houses often used in the tropics. The management of lighting regimes should take into consideration the following factors. The provision of light will encourage birds to consume more food and therefore grow faster. On the other hand intense lighting will encourage activity and increase metabolic rate and therefore cause birds to waste feed energy as heat. Strong light may also encourage feather pecking and cannibalism. One other point that should be taken into consideration is that broilers should be used to sudden darkness, otherwise a power cut, which causes the light to go out, may cause the birds to panic and smother. Probably the best compromise is dim lighting (a 60 watt red lamp for each 2 m^2) for twenty-three hours a day followed by one hour of darkness. If the birds are kept in red light during the growing phase and then light of a much shorter wavelength (e.g. blue light) is used when the birds are caught, this makes handling much easier. Birds that have been reared in red light are almost blind in blue light.

It is clear that chickens can be grown under light of any colour provided the correct intensities of illumination are used. Female broilers are more sensitive to brighter lights and perform best at intensities between 0.7 and 3.0 lux.

Temperature

The ideal temperature for broilers in the post brooding period is within the range 20–25°C. Research to date suggests that each degree of deviation from the optimum may result in a reduction of body weight at eight weeks of about 20 g per bird. Each degree rise in temperature is probably associated with a depression in cumulative food intake of about 50 g per bird.

Pre-slaughter treatment

If broilers are to reach the consumer in the best possible condition it is very important that the birds be treated in the correct way prior to slaughter. It is important that the coccidiostat should be withdrawn at four days prior to slaughter to remove the risk of residue in the carcass. It is also important that the birds should have the minimum amount of food in the intestine when they are killed. On the other hand if the birds are starved for too long they will lose weight. Probably the optimum time of starvation is four to six hours. Catching and handling of the roilers should be as careful as possible since rough handling will lead to bruises and downgraded carcasses.

Causes of downgrading

Carcasses will be downgraded if they are in poor condition due to disease (mycoplasma, leucosis, Marek's disease). They will be downgraded if they have breast blisters or abscesses. These may be caused by perching on hard surfaces and they will certainly be increased by the presence of caked litter.

Broilers in cages

This method of production is not recommended because of the effects of the system on carcass quality. Breast blisters may occur on up to 83% of carcasses. Another reason for avoiding this system of production is because of the danger of the birds breaking their legs in the cages.

Turkey production

Introduction

Turkeys can be marketed at several different weights:

- at twelve weeks (4.5 kg);
- at sixteen weeks (14.5 kg);
- up to thirty weeks (15 kg).

These large weight turkeys can be cut into joints and sold as turkey portions (breast, wings etc.). Turkey meat is also sold in pies and sausages.

The growing turkey has the advantage over the broiler as a meat producer in that it can attain much greater weights without exhibiting a serious decline in food efficiency. Thus they can be killed at ten to twenty-four weeks of age at weights of up to 14 kg without poor food conversion becoming a limiting factor. These heavy weights lead to birds with a killing-out percentage of 85% compared with a 70% figure for an eight week old broiler. Even at these very heavy weights turkeys have a very high protein content compared with broilers and other meat sources, but there is the disadvantage that the birds need to be fed high protein diets (Chapter 4). The high growth rates of present day turkeys are due in part to developments that have taken place in nutrition and breeding over the past thirty years, as can be seen by Table 6.1.

As a result of these changes the turkey has become one of the cheapest and most nutritious sources of animal protein in developed countries.

Table 6.1 Performance of white male and female turkeys to twenty weeks of age

Year	Liveweight at 20 weeks of age (kg)		Food conversion from 0–20 weeks of age
1957	Male	6.50	4.40*
	Female	4.20	5.40*
1973	Male	11.60	2.80*
	Female	8.00	3.10**
1988	Male	18.20**	2.40**
	Female	8.28**	3.75**

* Clayton and Nixey (1976)
** British United Turkeys Publicity (1988) (Big G breeding stock)

Management of growing turkeys

Young turkey poults are very difficult to rear, and if the level of management is indifferent, very heavy losses can be anticipated. In developing countries it is recommended that the young poults be reared in tiered brooders to three weeks of age although rearing under a traditional hover can be successful. Poults are difficult to encourage to feed and sprinkling of the food with a brightly yellowed food such as hard boiled egg yolk will prevent the occurrence of 'starveouts' (Fig 6.4). It is important also to cover any bedding with sacking (Fig 6.4) otherwise the poults may eat the bedding instead of the food provided. Poults are also liable to wander from the heat source and die of cold unless they are very closely confined. They need to be kept warm for about six weeks in the tropics but from three to six weeks of age turkeys can be reared in hay boxes and from then on in cattle yards and pole barns (Fig 6.5). Range rearing from eight weeks of age is possible on dry sandy soil but in wet conditions *Syngamus trachea* (the gape worm) can cause severe problems. Space requirements increase with age (Table 6.2).

Up to six weeks of age the turkeys should be fed on pellets in traditional tube feeders but once they are kept in pole barns large bulk feeders can be used (Fig 6.6). In order to maximise the efficiency of protein use, the protein level in the diet should be changed frequently, starting with a high level 30% pre-starter protein diet and reducing sequentially to a diet containing 14% protein for birds over eighteen weeks of age (Chapter 4). Because they have different nutritional requirements hens and stags should be reared separately from one day old.

During the rearing stages the traditional diseases of turkeys are black head (*Histomonas meleagridis*) and coccidiosis (*Eimeria* spp.). These are

Fig 6.4 *Turkey poults under traditional hover*

Fig 6.5 *Turkeys reared in straw (note the wire-covered 'A' frame in the corner to prevent the birds suffocating if they panic and crowd into the corners)*

Table 6.2 Space requirements per 100 birds

	Brooder space requirement for infra-red and other suspended heaters (m^2)	Total space requirement (m^2)
Growing turkeys		
0– 5 weeks	1.5	8
6–10 weeks	–	12
11–15 weeks	–	25
15 weeks to slaughter	–	40
Breeding turkeys		
Indoor management	–	150
Outdoor management	–	200

Fig 6.6 *Traditional type of turkeys reared in a hay barn and fed* ad libitum *from a bulk feeder*

easily controlled by prophylactic drugs in the food. In addition mycoplasma may cause a problem in intensively housed stocks – this disease is transmitted through the eggs, therefore eggs for human consumption should be purchased from mycoplasma-free stocks wherever possible. Fowl cholera (*Pasteurella multocida*) can occur – it is normally spread by wild birds.

As with broiler production the risk of disease will be reduced by periodic destocking and an 'all-in/all-out' system should be practised.

Turkeys are very susceptible to flavotoxin (almost as susceptible as ducks) and therefore groundnut meals should never be included in the diet of turkeys.

Temperature requirements

Turkeys grow more rapidly at temperatures lower than some found in developing countries. The recommended temperature post-brooding is 13–16°C. Temperatures below this will lead to an uneconomical increase in food consumption, whilst higher temperatures result in slower growth rates and fatter carcasses.

Processing and marketing

Processing is similar to chickens if the birds are under 6 kg in weight (up to sixteen weeks of age). Larger birds can be sold jointed. Males of 14 kg are particularly useful for this trade. They should reach this weight by the time they are twenty to twenty-four weeks of age.

Duck production

Management of growing ducks

Ducks are very nervous birds and they have a tendency to panic. They move around in flocks and if frightened when they are in a closed building, large numbers of them can be trampled underfoot in the ensuing panic. Because of this problem in the past ducks have been kept outside, even in temperate regions. There is no reason for this even when ducks are kept in large numbers. Water is not required for swimming or mating and provided the birds can immerse their heads in water they remain perfectly healthy.

With growing stock the objective is to maximise growth to killing age while minimising the food requirement. This objective is affected by the following factors:

- age at killing;
- food composition;
- food wastage;
- strain of duck.

Rearing ducklings from one day old

Rearing ducklings in the tropics should produce few problems since the young ducklings are very hardy. They should be reared in tiered brooders or conventional houses until they are two weeks of age, but thereafter they can be removed from any heat source. Since they suffer from few diseases they can be reared in a 'conveyor' system with the birds being moved to new pens every week until they reach the loading pen at eight weeks of age. Heavy meat type ducks need a fence of only 0.5 m to restrain them. They should be fed on pellets (young ducks tend to be very wasteful feeders). Ducklings of meat type breeds grow much more quickly than broilers reaching 3.5 kg in eight weeks but because of their high fat content they tend to have a poorer food conversion (3.5:1). However, the fat content of the carcass means that the birds require only a very low protein diet, in fact it is normally recommended that they be fed on a 16% protein diet from four weeks followed by one containing 14% protein. However, like turkeys they are very susceptible to flavotoxin and therefore should not be fed on diets containing groundnut as the protein source.

Selection for meat production in ducks

Ducks are water-fowl, and consequently contain a high proportion of fat, particularly subcutaneous fat. Selection programmes have therefore been carried out to increase the proportion of lean in the carcass, and particularly the proportion of breast meat. Work at Cherry Valley farm in Lincolnshire, UK has shown that selection for rapid growth rate, although it results in improved food conversion efficiency (a one day reduction in time to reach a fixed market weight results in a 1½% improvement in food conversion ratio) it also results in fatter carcasses because the proportion of breast meat increases as the duck approaches normal slaughter weight. Selection for an increased proportion of breast meat is successful but also results in a decreased proportion of lean meat in the legs so there is no overall improvement in lean meat production. On the other hand selecting for improved food conversion ratio increased the percentage of fat in the carcass.

7 Eggs and egg incubation

The avian egg

The shape of an avian egg is normally described as being an irregular ovoid (Fig 7.1). It can be divided into three main parts as follows:

Shell and shell membranes	12%
Albumen and chalazae	56%
Yolk	32%
	100%

Shell and shell membranes

The shell and shell membranes consist mainly of calcium carbonate and they form a protective surrounding for the soft inside of the egg. The shell is perforated by pores – thousands of minute intricate tunnels which allow moisture and gaseous exchange between the egg and particularly the embryo and the exterior. The shell itself is not uniform and consists of an outer layer and a spongy layer with the matrix towards the inside (Fig 7.1).

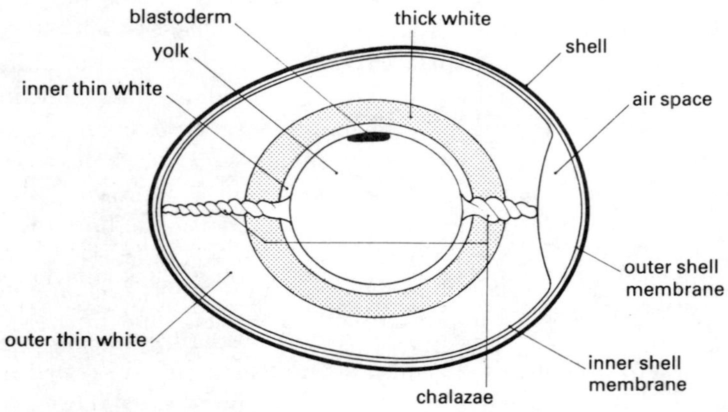

Fig 7.1 *Diagrammatic representation of an egg*

The shell membranes, of which there are two, divide at the crown end of the egg to form the airspace, which performs an important function in gaseous exchange, particularly at hatching time. It is into the airspace that the chick first breaks when it is on the point of hatching. If eggs are kept in a dry atmosphere particularly in a hot climate, moisture is lost from the egg and the airspace enlarges. This is why old eggs will float on water while new eggs will sink.

Both the shell and shell membrane fulfil a mainly protective function although in addition the shell also acts as a source of calcium for the developing egg. Because the shell is composed of almost pure calcium it is much denser than the rest of the egg. This fact is made use of when the specific gravity of the egg is used to determine shell thickness. It is not a precise measurement and it is only reliable for newly laid eggs. It is however a useful technique to use if large numbers of eggs have to be assessed without breaking. The way that this determination is made is to make up a series of brine solutions containing different concentrations of salt and therefore having different specific gravities. These brine solutions are placed sequentially in a row and baskets of eggs are dipped, starting with the solution with the lowest specific gravity. The eggs that float are removed and the operation repeated. The relationship between specific gravity and shell thickness is given in Table 7.1.

Table 7.1 Relation between specific gravity and shell thickness of freshly laid hens' eggs

Shell thickness (micrometre)	Specific gravity
355	1.086
244	1.059

The albumen and chalazae

The albumen located within the shell membranes is not a uniform substance. It is made up of four distinct layers (Fig 7.1). The thick white makes up about half the volume of the albumen and has layers of thin white on either side of it. The relative proportion of thick and thin white are not constant and they are for example, affected by disease, genotype of the bird and age of the egg. The longer the egg has been stored, the greater the proportion of thin white. This change is accelerated in hot countries. Thus if a newly laid egg is broken onto a flat surface it stands well away from the surface but an old egg will tend to spread because of its high content of thin albumen and will be close to the surface (Fig 5.18). An index of egg quality has been derived, the Haugh unit, which uses the height of the albumen as a measure of egg quality.

The chalazae are made up of coils of thick albumen, twisted in opposite directions, arising from the thick white layer at each end and terminating in the chalziferous layer, a very thin film of dense white that surrounds the vitelline membrane that encloses the yolk. The function of the chalazae is to keep the yolk located in the centre of the egg. They are not entirely effective and if eggs are not turned the yolk will settle to one side of the egg or the other. If this occurs during incubation it may result in the embryo not developing to the point of hatching.

The albumen acts as a source of moisture and protein for the developing embryo and at the same time absorbs products excreted by the embryo. Because it is composed largely of water it has a high specific heat and it therefore protects the embryo from large fluctuations of temperature.

The yolk

The yolk is formed in the ovary, not the oviduct. It is made up of layers of light and dark yolk. It functions as a food reserve and when the chick hatches it is absorbed within the chick's body. The presence of this food reserve enables the chicks to survive for several days after hatching without food and water. This long survival time means that newly hatched chicks can be transported over long distances, for example, from developed countries to form the basis of new poultry industries in developing countries.

On the surface of the yolk is the blastoderm (Fig 7.1). This is the seat of embryonic development. However, the blastoderm is present in all eggs and is not an indication that the egg is fertile.

Nutrient contents of the yolk and albumen of eggs

The yolk and albumen of hens' eggs differ considerably in their nutrient contents. The yolk contains far more protein than the albumen (Table 7.2).

Eggs produced by pullets that have just started to lay, produce eggs that contain a smaller proportion of yolk and a larger proportion of albumen. These eggs contain marginally less protein and fat and more water per unit weight than eggs produced by hens that have been laying for a longer period of time.

Table 7.2 Relative nutrient contents of yolk and albumen of hens' eggs

	Yolk %	Albumen %
Water	49.0	88.0
Protein	17.0	10.8
Fat	32.0	0.2
Ash	2.0	1.0
	100.0	100.0

Adapted from *Incubation and Hatchery Practices* (M.A.F.F. Bulletin, 148, 1977 p.2) courtesy of HMSO

Type of egg required for incubation

It is important that the correct eggs are selected for incubation if acceptable levels of hatchability are to be achieved. The example taken will be for the eggs of laying hens from modern hybrid stock. Eggs from other species require different treatment, these differences will be highlighted later.

All the eggs placed in the incubator should be of as uniform weight and size as is possible and have an average weight in the region of 55 g. Eggs which are more than 10% smaller than this weight may contain too small a yolk and therefore insufficient nutrients. However if there is a shortage of egg production it is permissible to use eggs as small as 48 g in order to fill the incubator. Eggs more than 10% larger may be double-yolked and if these are incubated they may prove to be infertile or produce deformed chicks. In the double-yolked eggs two chicks may develop at the same time and fuse together. Laying pullets in their first month of lay are particularly likely to produce double-yolked eggs. Incubation time is also affected by the size of egg and if the range of egg size incubated is too wide, hatching may take place over too long a period. The eggs produced by birds in the second half of their first laying year and during the second laying year normally have the highest hatchability. Birds in their third and following laying years often produce a greater proportion of eggs which are either infertile or the chicks fail to break out of the shell.

The eggs should be candled[1] to see whether they are sound. Eggs that are cracked should be discarded as cracked eggs lose water more quickly than non-cracked eggs and therefore these eggs will become desiccated

1 Candling is a process in which a bright light is shone through an egg to enable defects to be detected (Fig 7.5).

before the end of the incubation period and the eggs will fail to hatch. On candling, eggs containing blood spots and meat spots should also be discarded. Although these eggs will hatch normally, these two characteristics are inherited and therefore the discarding of eggs so affected will reduce the incidence of these in the next generation. It may also be possible to see breaks in the chalazae.

The eggs that are saved for incubation should be egg shaped, not long or round; these shapes are inherited and laying birds that produce these latter types of egg are not required for breeding purposes; therefore the removal of this type of egg will increase uniformity of shape of eggs produced by subsequent generations. The quality and thickness of the shell is also inherited. Eggs that are very thin shelled should be removed as should those with chalky and discoloured shells. The thickness of the shells can easily be determined by using specific gravity (page 149).

The management of the laying birds should be adequate to ensure that good quality fertile eggs are produced. The rations fed to breeding hens (as opposed to laying hens) need higher levels of certain materials (Chapter 4). Fertility is ensured by having adequate numbers of cocks per hen, a ratio of one cockerel per ten hens is recommended. It is better to have more than one cockerel with a group of hens to avoid the problem of preferential mating. If this happens certain hens are mated too frequently and others too infrequently.

The hens should be healthy. Hens that have been affected by infectious bronchitis produce eggs with distorted shells. A more serious problem is that hens can carry *Salmonella pullorum* in their ovaries. This will be passed onto the chicks causing bacillary white diarrhoea during their first week of life. With this disease a large percentage die. The presence of the disease can be detected by blood testing the breeding birds and eliminating the reactors. The air transport of day-old chicks from hens that have not been blood tested has disseminated diseases to several regions of the world.

The other chick diseases that are contracted by way of the egg are passed to the egg once it has been laid. These include omphalitis (mushy chick disease) – if this disease is contracted the chick goes rotten in the shell, (it is caused by a staphylcoccus) – chronic respiratory disease, and leucosis. The incidence of all these diseases can be reduced by washing the eggs in warm water containing a sanitiser. Eggs can be dipped in sodium pentachlorophenate as an additional precaution.

Treatment of eggs prior to incubation

The eggs should be stored for not more than seven days. They should be maintained at a temperature of 10–14°C and 75–85% relative humidity and should be stored broad end up. Fertility tends to decline by 1% per day after four days and 2% per day after ten days.

Incubation

Eggs can be incubated either naturally or artificially. The remarks below refer to the incubation of hens eggs, eggs of other species require different treatment regimes.

Natural incubation

Local breeds of poultry are good incubators and good mothers. They should be provided with a straw nest which has been disinfected before the beginning of incubation. Modern breeds of poultry, particularly those of laying strains have been selected for egg production, a process which has in part involved selection against broodiness. Consequently hens from these strains are not suitable for use as 'broody hens'.

Artificial incubation

Artificial incubation is carried out with incubators heated with fuels as diverse as paraffin, butane gas and electricity. Such methods of incubation, generally speaking, are not suitable for village poultry rearing. If artificial rearing is to be carried out it is probably better done in a controlled area where all the skills are available. This incubation unit could then supply day-old or older birds to the villagers. Incubators vary widely in their size and complexity (Table 7.3).

The incubators are managed in the following ways. In all of them eggs are packed in trays pointed end downwards.

Small incubators

These can be of several types. One type has a plastic top and can be used for demonstration purposes (Fig 7.2). A diagrammatic illustration of one heated by paraffin is shown in Fig 7.3. The operating temperatures of a small incubator are as follows.

- From the first to the nineteenth day the temperature is kept at 38–39°C and the humidity at 60–65%.
- From the twentieth to the twenty-first day the temperature is increased to 39–40°C and humidity to 70–75%.
- The eggs are also turned four times a day through ninety degrees for the first twelve days. From that time until hatching they are not turned. For more information see M.A.F.F. *Incubation and Hatching Practices* (1977).

Table 7.3 Types and sizes of incubator

Type	Capacity	Fuel and management	User
Small incubator	50–300 eggs	Paraffin, butane gas or electricity; natural ventilation and eggs turned by hand. (They are difficult to use and often function badly.)	Not recommended
Incubators with ventilation	3 000–30 000 eggs	Electricity; fan ventilation	Large producers
Air conditioned incubators	very large capacity	Electricity; air conditioned	Very large operations

Mammoth incubators

These machines should be operated according to the manufacturer's instructions. The machines have separate incubators and hatching compartments. Eggs are kept in the hatching compartment from the first to the nineteenth day at a constant temperature of 37.8°C and a humidity of 60% (wet bulb temperature 26°C). On day nineteen they are transferred to the hatching compartment at the same temperature (or possibly 1°C lower to avoid overheating) but with the humidity raised to 80%. The eggs are turned automatically up to eight times a day until the eleventh day.

At all times the eggs should be adequately ventilated. Ventilation is a way to regulate temperature and humidity. It also serves to bring oxygen to the chick and remove excess carbon dioxide. The removal of air and moisture from the eggs can be an expensive business because heat is removed at the same time. This applies less to tropical countries than to temperate climates. The rate of ventilation needs to be increased as the chicks grow because as they grow so does their rate of metabolism. Therefore, air exchange increases from only one air change every hour when the chicks are first put into the incubator to twelve complete changes an hour just prior to incubation.

Incubator hygiene

Eggs are very suitable sources of nutrition for micro-organisms and the temperature at which eggs are incubated is one which is highly suitable for many types of bacteria and viruses.

Fig 7.2 *Chicks hatching in a simple plastic-topped incubator*

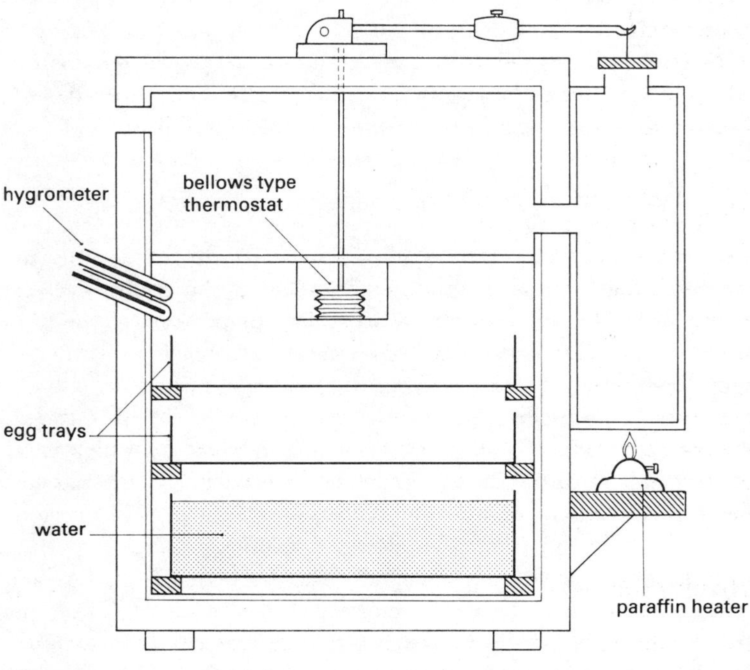

Fig 7.3 *Diagrammatic representation of a small incubator* (Chris McKinnell)

The keeping of the incubator unit, the eggs and the operators as clean as possible is essential if disease outbreaks are to be avoided.

The premises in which the incubators are housed are normally designed to allow for sequential cleaning. All rooms and equipment must be cleaned and disinfected regularly. The egg handling room should be cleaned thoroughly and disinfected once a week. The incubators should be thoroughly cleaned and disinfected after each batch of eggs have been hatched and the chicks removed. Incubators can be fumigated with certain types of fumigants, for example with formaldehyde gas, when they have eggs or chicks *in situ*. After each use, all other equipment and transport should be disinfected.

Only eggs from blood tested stock should be brought onto the incubator premises. They should be clean and carried in clean containers. Before they are put into the incubators they should be disinfected either by dipping in a sanitiser or fumigated. Disinfectants suitable for this purpose include sodium pentachlorophenate at 0.5% and quarternary ammonium compounds. All should be used according to the manufacturer's instructions.

The personnel who handle the chicks should not come into contact with those who look after the adult poultry. They should wear overalls and head coverings and disinfect their hands and feet before entering the buildings. When they have worked in the incubator they should change their clothes and disinfect their hands before entering the area where the eggs are received.

Critical periods in the incubation of eggs

The death of chicks in the shell does not occur at a uniform rate throughout the incubation period and there are a number of critical periods. 25% of the total mortality takes place during the third to fifth day of incubation (Fig 7.4). This critical period is thought to occur because at this time the developing chick goes through a number of physiological changes. For example at this stage the carbohydrate stores of the chick are used up and the chick then has to rely on fat and protein as sources of energy. These are rather more difficult to metabolize by the chick. It is also during this period that the respiratory system of the chick is evolving. The second major critical period occurs after the chick has been incubated for eighteen to twenty days and half the mortality occurs during this period. Death at this late period is associated with the chick moving around in its shell to get into the right position for hatching. If the chick gets into the wrong position the chick will die. This condition is made

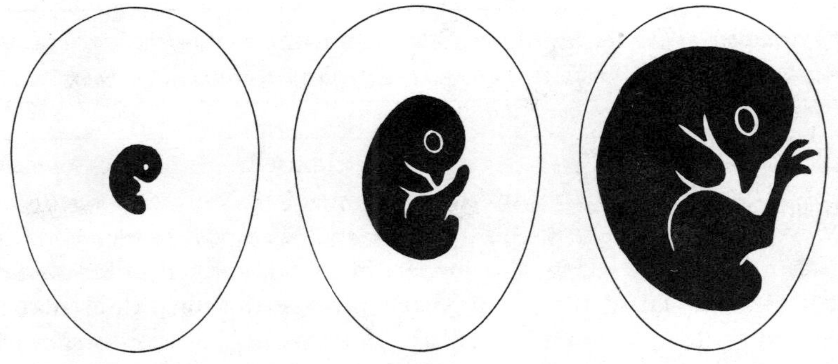

Fig 7.4 *Stages in the development of the chick embryo* (Chris McKinnell)

worse if the eggs have not been regularly turned due to malfunctioning of the incubator caused for example by frequent failure of the supply of electricity.

Causes of mortality

The number of eggs which are infertile or only develop partially can vary to a very large degree depending on circumstances. If for example the brooding temperature or relative humidity is too high or too low the percentage hatch will be decreased, the amount of decrease depending on how far the temperature is from the ideal. If the temperature is too high the chicks will hatch early, if too low the chicks will hatch late. Similarly if the eggs are turned an insufficient number of times fertility may be reduced and the number of dead-in-shell increased as they will have been stored for far too long or at the wrong temperature prior to incubation.

Lethal genes may be a problem especially in inbred populations associated with the breeding of hybrid populations but should not be a problem in normal outbred populations. Disease should not be a problem if adequate hygiene is practised.

Fumigation of eggs

Eggs should be fumigated as soon as possible after they have been laid. Use 114 g of formaldehyde with 85 g of potassium permanganate for every 3 m^2 of space to be fumigated. The fumigation should last for twenty minutes.

- *Caution 1 The reaction between the two chemicals is violent.*
- *Caution 2 Do not fumigate eggs twenty-four to seventy-two hours after setting them in the incubator.*

Candling of eggs

Eggs can be candled (Fig 7.5) for fertility at day seven of incubation. Since this is a fairly labour intensive process this operation may be dispensed with in a modern poultry unit but in developing countries the eggs could be removed and used for human consumption. However such eggs are not well liked by local populations and are best reserved for commercial use, e.g. in the production of cakes. If the egg is 'clear' it is almost certainly infertile but if it contains signs of development, i.e. the development of blood vessels, it is almost certainly fertile. Normally in the region of 15% of the eggs would be classified as clear. Of these about 12% would be caused by infertility and 3% by early deaths of the embryo. Eggs can also be tested on day eighteen but this is rarely carried out now. If too many eggs are infertile this may mean that faulty AI techniques were used or too few males were used in relation to the number of females. If there are too many dead-in-shell this could be caused by irregular or insufficient turning of the eggs, genetic problems with the breeding stock or possibly an infection with salmonella.

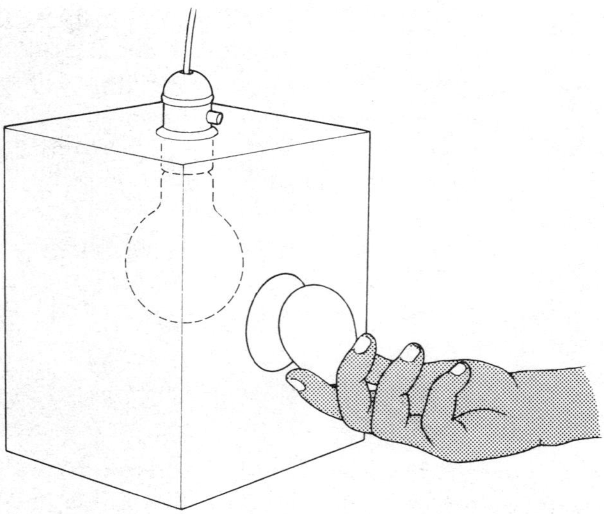

Fig 7.5 *A simple candling device* (Chris McKinnell)

Hatching

The eggs are moved to the hatching compartment (Fig 7.7) at day eighteen where the humidity is rather higher than in the incubator (Fig 7.6). The chick draws the yolk sac with the body at hatching and this acts as an additional source of food for the first week after hatching. It is the presence of this yolk sac that enables chicks to be transported over a distance involving several days travel without the need to provide them with food and water. It also means that chicks should not be used in nutrition experiments for the first week after hatching because the absorption of nutrients from the yolk sac will influence the results obtained.

At hatching the chick jerks its head by reflex and breaks its way through the allentois and starts to use its lungs for the first time. The changeover from absorption of oxygen through the allentois to lung respiration supplies the chick with more oxygen and as a result it becomes more active and is able to break a small hole in the shell. This breakthrough which is known as 'pipping' is achieved with the aid of a small egg 'tooth' at the end of the beak. Once the chick has broken through, it braces its legs against the shell and turns around in the shell, pecking as it goes. As a result it neatly removes the top of the egg.

Fig 7.6 *Eggs in incubating compartments*

Fig 7.7 *Eggs in hatching compartments*

Sexing of chicks

The sex of a chick may be determined at hatching time by everting its vent and looking for the presence or absence of a tiny rudimentary copulatory organ on the inside rim. The best time to carry out this operation is five to fifteen hours after hatching when the chicks have dried off and have the strength to be handled but are still very pliable. This operation cannot be carried out by a normal farm worker as it is a very skilled operation. It is normally carried out by teams of specialised 'sexers'. These people should be capable of correctly sexing 500 chicks an hour.

Sexing older birds

Apart from vent sexing it is possible to determine the sex of birds in other ways such as by the use of sex linked colour, auto-sexing, barred breeds and auto-sexing feather development (Chapter 2).

It is possible to determine the sex of all chicks at ten to twelve weeks by reference to the growth of the tail feathers. These grow much longer in the cockerel. However twelve weeks is far too long to wait for birds of laying and broiler strains and therefore other techniques have to be used.

Output of chicks from incubation

It is self evident that fewer than 100 chicks will be produced from each 100 eggs laid but the output is much lower than perhaps expected especially when laying pullets are being produced because in this case only the females are required and the males are discarded.

Some idea of the likely output is given from the following figures:

100 eggs laid
 10 discarded
—
 90 remaining
 10 clear
 10 dead in shell
—
 70 remaining
 4 culled
—
 66 remaining
 33 cocks
—
 33 pullets remaining
—

Therefore about 33 pullets are produced from each 100 eggs laid.

Incubation of turkey eggs

Turkey eggs have an incubation period of twenty-eight days. The eggs are considerably heavier than hen eggs, weighing from 70–100 grams depending on the size of the breed of turkey. Special turkey incubators should be used since those used for chickens have trays that are too small. The incubation regime required is similar to that used by chickens for the first twenty-one days (38°C, 60% relative humidity) but the temperature should be dropped to 36°C and the relative humidity raised to 75% for the last week in the incubator.

Incubation of duck eggs

Duck eggs have an incubation period of twenty-eight days (Muscovy ducks: 33–35 days). Their incubation is a similar process to that of chicken eggs except that a slightly higher relative humidity is required (37.5°C and 75% relative humidity) during the first twenty-five days of incubation. A temperature of 36.9°C is required during the hatching period and a humidity of 90% for the last three days. During this latter period the eggs should be sprayed with warm water. Duck eggs should be turned more frequently than hens' eggs (once an hour).

Traditional incubation

The artificial incubation of eggs has been practised for several thousand years in the Nile Delta and the practice of incubating the eggs of native Fayoumi birds in commercial incubators is still common today. The eggs are incubated in large insulated purpose-built buildings and additional heat may be provided by bean straw which burns slowly.

A similar practice is used in Bali, Indonesia to incubate duck eggs although this is a much more recent development. In this system the eggs are put in a basket made of bamboo which is lined with a sack to keep the eggs warm. The basket which is 90 cm high and has a diameter of 50 cm holds approximately 1000 eggs. The eggs are first warmed in the sun for three hours and are then packed into the baskets between layers of dry sun-warmed unhulled rice. Each layer is separated by a thick piece of cloth. The eggs are kept in this basket for three days and start to incubate due to the warmth of the eggs and the rice. The eggs are turned seven times a day. After fifteen days of incubation the eggs are moved to hatching tables where they are covered with hessian sacks, turned seven times a day and stay on this table until they hatch. With this system in Bali, 80% of the eggs that are set, hatch.

8 Disease and disease prevention

Introduction

It is always better to prevent a disease than to cure it. Sick birds will always be unproductive, so birds should always be kept as free as possible from disease in order to ensure both maximum profitability and productivity. The constraints applied by disease are particularly important when birds are housed under intensive conditions with birds in close proximity to each other, thus affording the maximum opportunity for diseases to spread from one bird to another. However, even under extensive conditions birds can easily spread disease because they are normally housed close together at night in order to protect them from thieves and carnivorous animals. If birds of different ages and of different species are housed in this way at night the spread of disease is encouraged.

Many of the diseases and parasites that cause problems are common to the tropics, subtropics and temperate zones. However, in addition there are diseases such as fowl pox which are not found in temperate zones. It is important therefore that governments ensure that diseases are not imported from developed countries by making certain that disease-free breeding stock is imported. It is important that parent flock vaccination against avian encephalomyelitis, Newcastle disease, Gumboro disease, EDS 76 and infectious bronchitis is carried out and that chicks are vaccinated against Marek's at hatching and spray vaccinated for infectious bronchitis after hatching.

What, therefore, are the precautions that can be taken by the poultry keeper to ensure that the introduction and spread of disease is minimised?

This chapter attempts to outline various methods of disease control including house siting, house hygiene and house management. In addition the use of vaccines and prophylaxis will also be discussed.

The siting and management of poultry houses

The siting and management of poultry houses is important in any disease control programme. It is desirable to ensure that birds of different ages are not kept on the same farm and, at the very least, not kept in close proximity. If birds of one age only are kept on one site there is far less chance of disease spreading from older birds to younger ones. This can be achieved if the poultry keeper follows the all-in/all-out system. In this system whether the birds are used for laying or for broilers all the birds are brought onto the site at the same time and at the same age. The result of this system is that there is only one opportunity for diseases to be brought onto the site, the birds will be more likely to respond to a programme of vaccination and it is much simpler to administer a disease prevention programme. At the end of the period, whether it is at the end of the growing period for broilers or at the end of the laying season for laying birds, all livestock are removed from the property, the manure is cleaned out, the houses thoroughly disinfected and a gap of at least two weeks is left between the selling of one batch of birds and the bringing in of a new batch.

It is very tempting under intensive management conditions to replace dead birds with others of a similar age. The farmer sees this as a way of using an under-utilised resource such as an empty battery cage. This procedure should, however, be avoided because the risks of introducing disease far outweigh the extra benefit achieved by filling an empty productive unit.

By a similar token it is also advisable to remove sick birds from a house to avoid them becoming a source of infection for other birds. Sick birds are very obvious to the trained observer. In the case of adult laying birds, the obvious sign of sickness is that they stop producing eggs. Other signs are that the birds are in a poor condition, the feathers look soiled and combs can become small and discoloured. A bird may try to isolate itself from the rest of the flock. Younger birds may have watery eyes and traces of diarrhoea around their vents and they may walk with an unusual step, and they are normally fairly inactive. If there is a sudden decrease in total consumption of food it can be expected that something is wrong with the flock; again, if mortality exceeds 1% in any one-week period this is an indication that something may be seriously wrong. However, even in a well managed flock it is expected that at least 1% of that flock will die each month. When dead or dying birds are removed from the flock they should be examined post-mortem to see if the cause of the illness can be determined and then the remains should be buried. Hands and utensils should be thoroughly cleaned before any other job is undertaken. If the level of mortality is above what is deemed to be normal, veterinary advice should be sought immediately.

Table 8.1 Important diseases controlled by vaccination

Name	Means of transmission	Causative organism	Signs	Prophylaxis	Treatment
Newcastle disease	Bird to bird by droplets in air.	Virus	Paralysis. Difficulty in breathing. Greenish diarrhoea.	Vaccination either dead or live vaccine.	None.
Infectious bronchitis	Bird to bird.	Virus	Respiratory problems. Large decrease in egg production.	Vaccination.	Antibiotics to control secondary infections.
Fowl cholera or pasteurellosis	Through water and food to nose and mouth.	Bacterium of *Pasteurella* spp.	Severe diarrhoea. Blue combs and wattles. Reduction in food intake.	Annual vaccination with live vaccine.	Remove, slaughter and destroy infected birds. Clean infected premises.
Marek's disease and leukosis	Bird to bird.	Virus	Affects birds 12–24 weeks old. Causes paralysis and death of 10–30% of flock.	Vaccination. Isolation.	None.
Fowl pox	Mosquitoes and other biting insects and through damaged skin.	Virus	Scabs on the comb, wattles, eyelids. Death.	Vaccination. Isolation.	Cull clinically affected birds. Vaccinate uninfected ones.
Gumboro disease (infectious bursitis)	By direct contact especially between young birds.	Virus	Prostration of birds. Diarrhoea.	Isolation. Vaccines are available.	None.
Infectious hepatitis (duck virus)	Bird to bird from faeces.	Virus	Young ducks stop eating, fall over, legs stretched out and mortality very high.	Destock farm, clean, disinfect. A vaccine is available.	None. NB Do not treat with nitrofurans because these are toxic to ducks.

Control by vaccination

Many of the poultry diseases are found in the tropics, subtropics and temperate zones. However, in addition in the tropics there are diseases, such as fowl pox which are not usual in temperate zones but are commonly found in many tropical countries. Before starting any programme of vaccination it is important to know what diseases are found within the area or the country as there is no point in wasting money on vaccines for diseases which are not present. On the other hand it is vital to vaccinate against those diseases which are prevalent and for which a vaccine is available (Table 8.1).

It is important to remember that vaccines are made from the disease which you are trying to prevent. They should always be stored according to directions so that they do not lose their viability; many of them need refrigeration and expiry dates should be checked. It is possible to keep vaccines in a useable condition by maintaining what is known as the 'cold chain'. The vaccinator can take the vaccine to the village in an insulated box containing ice which is shaded from direct sunlight. Alternatively the vaccine is sold to people who have been shown how to vaccinate and who have their birds cooped up at the time of purchase so that they can return straight home and begin to vaccinate immediately. Vaccines should always be mixed away from poultry and no spilling on or near chickens should occur. The correct dose of the vaccine must be administered. Trying to make the vaccine cover too many birds is always a false economy. Vaccines should never be kept overnight as they may lose their effectiveness. The directions need to be read carefully and the use of clean, well-rinsed, sterilised equipment is necessary. Either thoroughly sterilise equipment after use or, if possible, dispose of it and use new equipment. Also ensure that clothes are changed and shoes disinfected before visiting other chickens. Methods of vaccination include:

- the use of an eye dropper to place the vaccine in the eye;
- the use of a vaccine-soaked cotton swab or brush to rub the vaccine into the place where a few feathers have been removed;
- using the stab method for pox (equipment comes with packet);
- in the case of some vaccines — administering them in the drinking water.

Control by prophylaxis

A second way of preventing disease is by the use of good management associated with prophylactics such as coccidiostats and sulpha drugs (Table 8.2). There are a number of different types of coccidiostats which appear under many different names. These should be used sequentially

Table 8.2 Important diseases controlled by prophylaxis

Name	Means of transmission	Causative organism	Symptoms	Prophylaxis	Treatment
Coccidiosis	From the droppings of infected birds.	Protozoa of *Eimeria* sp.	Watery and bloody diarrhoea. High death rate.	Separation of adults and young. Feed coccidiostat.	Treat with a sulphonamide, pyrimidine or amprolium in the drinking water.
Black head (turkeys)	Bird to bird by way of an intermediate host (a parasitic nematode).	Protozoa in the caecum (*Histomonas meleagridis*).	Yellow-coloured diarrhoea.	Phenothiazine, Dimetridazole or Ipronidazole.	Dimetridazole or Ipronidazole.
Pullorum disease (bacillary white diarrhoea)	From hen to chick through eggs.	A salmonella bacterium.	High death rate in chicks. White diarrhoea.	Blood test adults to to remove carriers.	Use of sulpha drugs such as Furazolidone

so that the local species of the parasite does not establish resistance to any particular one. Do not mix one coccidiostat with another since it may be harmful, and always follow the manufacturer's directions. All should be mixed in the food but should be withdrawn four days before the birds are slaughtered so that the coccidiostat is not carried over into the flesh for human consumption. Amprolium may be used to treat birds when an outbreak of coccidiosis occurs in a flock.

The use of sulpha drugs in food or water will eliminate infectious coryza which is caused by a bacterium and spread from chronic or acutely infected birds to healthy ones. If the flock is treated in this way and there are no other complications the disease will be removed within two weeks.

Diseases that can be prevented or controlled by vaccination

Newcastle disease or fowl pest

Newcastle disease is a disease caused by a virus which is found worldwide in chicken flocks and turkeys and it is endemic in many countries. It is very contagious spreading from bird to bird in droplets and moisture in the air. The removal of the carcasses of birds that have died of this disease by carnivores and birds of prey can spread this disease. Cases have been reported of its spread from sick pigeons to laying fowls by way of contaminated layers' mash. When it strikes young chickens the mortality rate may be as high as 100%. In adult birds mortality rates are lower but an attack will have a very marked effect on egg production. Outbreaks of the disease often occur at specific times of the year associated with the migration of wild birds and changes in climatic conditions. The most obvious symptom is a sudden high number of deaths.

Other symptoms of the disease are paralysis, difficulty in breathing, greenish diarrhoea and a discharge through the beak. The crop may also be distended, the head may droop, the bird shows drowsiness and ultimately coma and death will ensue. It is necessary to confirm the diagnosis in the laboratory.

Vaccination

To prevent the spread of this disease all birds, including broilers, should be vaccinated against it. There are two types of vaccine in use: a dead vaccine and a live vaccine. The live vaccine should precede the use of a dead vaccine. When used, the latter is most useful in commercial layers and in breeding flocks. The live vaccine is given in the drinking water. Two applications are needed, the first is given in the first week post-hatching and the second about two months later. The live vaccine gives better protection but is more risky in its use in that it may result in the spread of disease to other flocks. Timing of the vaccination programme is important if seasonal outbreaks are to be minimised. If chronic respiration disease is present vaccination against Newcastle disease may precipitate pneumonia (respiratory disease). Antibodies against chronic respiratory disease may be incorporated in the Newcastle vaccine.

Treatment
There is no treatment for this disease.

Post-mortem
The obvious signs on post-mortem are mucous in the trachea and possibly yellow and thickened air sacs with haemorrhages in the proventriculus.

Infectious bronchitis

Infectious bronchitis is a very contagious disease which is particularly serious to immature chickens. It causes respiratory troubles, and high mortality rates can result from secondary infections caused by bacteria. The virus also causes lesions in the oviduct and as a result the hens produce abnormal eggs and a very large drop in egg production can also occur. In chicks mortality may be from 5–60% but in mature birds only a very small number die, usually less than 2%.

Vaccination
A live attenuated vaccine is available. The best method of control is the use of a mild live vaccine followed by an oil adjuvanted[1] inactivated vaccine. Vaccination for infectious bronchitis is often given to the birds at the same time as the Newcastle disease vaccine is applied. At the same time all infected birds should be disposed of and houses and equipment of the infected birds thoroughly disinfected.

Treatment
There is no specific treatment for this disease but antibiotics can be given to reduce the effect of associated disease.

Post-mortem
It is difficult to determine death from infectious bronchitis as there are many other diseases which affect the respiratory apparatus — chronic respiratory disease, lack of vitamin A and Newcastle disease.

1 assisted

Fowl pox

This is a virus disease which is very prevalent especially among extensively managed flocks. The virus is very hardy and can remain viable in scabs for up to ten years. It is mechanically transmitted by mosquitoes and other blood sucking arthropods.

Vaccination
Annual vaccination of all birds should be carried out usually at the same time as the Newcastle disease vaccination.

Treatment
There is no recommended treatment as topical application of medicine usually spreads the virus.

Fowl cholera or pasteurellosis

This is a bacterial disease which is both infectious and contagious and which can cause a high level of mortality in both chickens and in ducks. The symptoms of the disease are severe greenish-yellow diarrhoea, difficulty in breathing, enlarged hot swollen combs and wattles and discharges from the nasal passages. There is a general lack of energy and unwillingness to eat.

Vaccination
An annual vaccination of all birds with a live vaccine may prevent the disease.

Treatment
Sick birds can be treated with intramuscular injections of terramycin or streptomycin, while the birds that have escaped the infection should have antibiotics or sulphonimides added to the drinking water for several days. Infected birds should be removed, the litter changed and equipment and houses thoroughly disinfected.

Post-mortem
Findings would include congested viscera and septicaemia.

Marek's disease

This is a virus disease spread in the cells sloughed off from chicken feathers. This is a major disease of importance to species of egg laying birds because Marek's disease normally affects layers within two or three weeks of the birds coming into lay. Mortality rates can be 1% or 2% per day for two to three weeks and this results in a considerable loss in the productive potential of the flock. Some birds are mildly to completely paralysed and will lie on their sides and slowly starve to death.

Vaccination

A live vaccine for this disease given at one day old has been available for a number of years and achieves a high degree of control. However, recently there has been increasing evidence of breakdown in vaccinated fowls. This has probably been due to the emergence of new, highly virulent strains of the virus. These can be controlled with a new vaccine developed in the USA.

Treatment

There is no known treatment for this disease.

Post-mortem

On inspection it will be seen that the femoral nerve has 'knots' or swellings in it (compare with a healthy bird).

Diseases controlled by prophylaxis

Coccidiosis

Coccidiosis is a disease that has become more important with the advent of intensive management. It is caused by nine distinct species of protozoan parasite and it is spread to chickens, particularly young chickens, through contaminated food, water or litter. It has not been possible to eliminate this infestation because the sporulated oocyst of the protozoan excreted in the faeces is highly resistant to chemical disinfectants. A small number of birds remain carriers throughout their lives and continue to excrete small numbers of oocysts. Each species attacks a different portion of the intestines or caeca and it is particularly prevalent in chickens that are kept intensively which have the maximum possibility of taking up the disease from their fellow chickens. The disease can be prevented in one of three ways. Firstly, when the birds are kept extensively, the natural immunity of a flock can be allowed to build up without the administration of any drugs unless there is an outbreak. This approach is only safe to adopt when there are low levels of coccidiosis, but under very dry conditions there may not be enough infection around for the birds to develop immunity. If the environment is very dry it is advisable to sprinkle the litter with water at weekly intervals to ensure that adequate exposure is achieved.

In the case of intensively managed flocks, complete prevention is advised by the feeding of a coccidiostat throughout the bird's life. This will prevent the bird from contracting coccidiosis at all, and is useful for the production of broilers but the coccidiostat should not be included in the food for four days before the birds are slaughtered. If this is not done people may consume meat which contains drugs whose side effects are unknown but potentially serious.

The third possibility is to try and develop immunity and combine this with treatment. With this method, which is suitable for laying birds, use a low level of prophylaxis with a coccidiostat for the first fifteen weeks which will allow the birds to develop some immunity and only treat the birds at a therapeutic level if outbreaks occur.

This may be achieved in a more sophisticated way by vaccination by continual dosing with small numbers of oocysts with the intention of building up immunity. This can be achieved by means of the water or by the food. Live oocyte vaccines are commercially available in the USA to enable this approach to be taken.

Chronic respiratory disease (also known as avian mycoplasma)

This is an infectious disease of poultry caused by *Mycoplasma* spp. Symptoms are a rash in the throats of poultry and death by asphyxia. The respiratory tracts and air sacs of the lungs are normally inflamed and sometimes filled with exudate. The disease is normally a secondary infection following either parasitic or microbial infections or possibly after some stress factor such as vaccination.

Control

All infected birds should be destroyed and contaminated equipment disinfected along with the buildings. All replacement birds should be put in quarantine. The disease can be egg transmitted and a vaccine is available for use with breeding flocks to prevent transmission through eggs. Replacements should be obtained from mycoplasma-free stock (blood tested stock).

Treatment

It is possible to treat birds with antibiotics of the tetracycline groups (erythromycin and tylosine) but these will not give a complete cure.

Fowl typhoid and pullorum disease

These diseases are still found in many parts of the world. They are caused by salmonella bacteria. Some *Salmonella* spp. which affect humans also produce similar symptoms to both diseases. The bacterium can survive in fowl carcasses and outside the body of the fowl and it is spread by means of the eggs and faeces of infected birds and by visitors to the plant. The symptoms of the disease are drooping wings, high mortality rates and diarrhoea. In chicks infected with pullorum disease the diarrhoea is white and sticky and this leads to its common name of bacillary white diarrhoea. Over 50% of the chicks may die.

Control

The disease can be controlled by hygiene measures in management, such as trying to isolate individual flocks (fencing), keeping scavenging dogs away from flocks, by blood testing and removal of reactors and by vaccination. The best way to avoid the disease is to buy only chicks certified to be free from salmonella. Unfortunately it is not unknown for the health certificates that accompany day-old chicks imported by air to be at variance with the facts. It is therefore important that authorities ensure that the health certificates have been prepared and signed by a reliable individual. Vaccination (live, attenuated) is not entirely satisfactory and will only give protection over a few months. The use of vaccination may interfere with the blood test which should be the main method of control.

Treatment

This is by the use of sulpha drugs and Furazolidone (nitrofurans).

Post-mortem

There may be signs of septicaemia even in birds killed for post-mortem examination and a greeny-bronze liver; abnormal lesions may also be found.

Duck virus hepatitis

This disease causes high mortality rates in ducks of one week of age. Affected birds stop eating and fall over onto their sides. Their eyes are partially closed and the birds die suddenly with their heads up and their legs stretched behind them. All ducklings in the group become infected and mortality rates can be very high.

Control

A live vaccine is available. When the disease occurs the farm should be destocked, thoroughly cleaned and disinfected. The virus may persist for long periods in the absence of good hygiene.

Treatment

There is no satisfactory treatment; nitrofurans should not be used as they are extremely toxic for ducks.

Duck virus enteritis

This disease is found in wild water-fowl and therefore sporadic outbreaks in domestic stock are not uncommon. The disease can be controlled by vaccination.

Internal and external parasites

Parasites, both internal and external, are common in the tropics (Table 8.3). Both will weaken the bird and in certain cases may be a secondary host for other diseases.

Internal parasites

Helminth infection is extremely common in poultry especially when they are kept under intensive conditions. Of these the most important are roundworms (nematodes) and tapeworms (cestodes). They are spread from one chicken to another through the faeces or through a variety of intermediate hosts including snails, earthworms, flies and other insects. Some, which are found in the windpipe (trachea), and others, which are found in the intestine, will result in a decrease in food intake, liveweight gain, and egg production. Because they usually cause subclinical diseases which are rarely lethal it is difficult to calculate the cost of them to the owner.

Control
It is impossible to prevent the complete uptake of worm eggs by chickens but the numbers can be reduced, and the number of worms in the intestines depends on the number of eggs that have been taken up by the bird. Litter must be replaced between batches and wild birds need to be kept out of the house.

Treatment
Ascaridia spp., *Capillaria* spp. and *Heterakis* sp. can all be treated with piperazine in the water.

External parasites

These are of three types:
- lice;
- mites;
- ticks.

The most common parasites are the lice.

Table 8.3 Principal parasites affecting poultry

Name	Means of transmission	Causative organism	Symptoms	Prophylaxis	Treatment
Internal parasites					
Ascaridia spp.	Infected runs.	Nematodes 3.5–7.5 cm long in small intestine.	Poor health. Diarrhoea.	Move chicks to clean ground. Disinfect houses. Separate adults and chicks.	Piperazine in food.
Capillaria spp.	Infected runs.	Nematodes 1.0–1.5 cm long in oesophagus.	Lack of appetite. Diarrhoea bloody.	Move chicks to clean ground. Disinfect houses.	Piperazine, haloxon or methyridine.
Heterakis gallinarum	Infected runs.	Nematodes 0.7–1.2 cm long in the caecum.	Limited but linked to parasites that cause histomoniosis.	Kill intermediate hosts. Antibiotics in food.	Piperazine. The antibiotic Hygromycin B
Syngamus trachea	Infected runs.	Nematodes in the trachea.	Gasping for breath. Death by suffocation.	Thiabendazole in diet. Move to clean runs.	Thiabendazole in diet.
External parasites					
Ticks	From infected buildings and birds.	Blood-sucking parasites.	Loss of weight. Loss of appetite.	Keep equipment clean. Avoid use of wood.	Nicotine sulphate on perches. Dust birds with malathion.
Mites	From infected buildings and birds.	Blood-sucking parasites (smaller than ticks).	Loss of weight. Loss of feathers.	Keep equipment clean. Avoid use of wood.	As for ticks except for scaly leg mite. Birds infected with this mite should be culled.
Lice	From infected birds.	Parasites that that spend life cycle on birds.	Ill health. Skin irritation	Nicotine sulphate on perches.	Dust birds with malathion.

Lice

Lice differ from mites and ticks in that they spend their entire life cycle on the body of the bird. The symptoms of lice infestation are those of general ill health, slow weight gain and decreased egg production associated with skin irritation, scabs and ruffled feathers.

Control

The nests should be kept clean and if the birds roost at night, perches should be painted with nicotine sulphate, and creosote placed in the cracks and joints of the roost.

Treatment

There are a number of substances which will control lice on birds including malathion and the pyrethrins and these can be dusted onto the birds. In the case of deep litter systems, the litter and all equipment except the feeders and waterers should be sprayed and dusted.

Ticks

Ticks are blood sucking creatures which have various forms mainly from flat to egg-shaped to disc-shaped. In addition to being blood sucking they can also carry diseases such as spirochetosis and tick fever. They are normally found only in hot dry areas and they spend part of their life cycle in cracks in walls and roofs and in other equipment made of wood and they are very hard to locate. The symptoms of tick infection are loss of appetite, loss of weight and possibly anaemia.

Treatment

Nests should be kept clean and if roosts are used they should be painted with nicotine sulphate. Spraying should be carried out once every two weeks. A thatched roof can be a good haven for ticks and it may be necessary to replace it if the infestation is bad. If the houses are old it may be worthwhile burning them down and starting afresh. In West Africa such roofing material is not recommended for poultry houses.

Prevention

Avoid the use of wood in equipment, such as feeders and nesting boxes.

Mites

Mites are smaller than ticks and almost invisible. Unlike ticks some stay on birds continuously whilst others only attack at night and hide in the woodwork and litter during the day.

Symptoms

The symptoms depend on the type of mite, for example scaly leg mite makes the birds lame and the legs have large swellings which are scaly and encrusted. On the other hand, red mites, which only attack at night, cause loss of weight and loss of egg production and may lead to anaemia. Depluming mites attack the base of the feathers and birds pull out feathers to relieve itching, hence their name.

Treatment

There is no treatment for the scaly leg mite except for culling infected birds. For other mites the treatment is the same as for lice and ticks.

Prevention

The same as for ticks.

Mycotoxicosis

The most important toxin in this category is the aflatoxin produced by the fungus *Aspergillus flavus*. The chief source of mycotoxins are stored grains of cereal or oil seeds, particularly when the seed is damaged and/or contains more than 11% moisture. The toxins are highly resistant and may be present in food when all evidence of the mould has disappeared. Ducks and turkeys are particularly vulnerable to this toxin and levels as low as 1 μg of aflatoxin per gram of food is lethal for ducks and turkeys. Chickens are less vulnerable and are only moderately affected by levels as high as 4 μg per gram of food.

Under tropical conditions mould growth is a serious problem and most batches of groundnuts are affected and should not be fed to turkeys or ducks. If no checking mechanism is available assume that all samples are infected and limit the intake of groundnuts by chickens to 10% of the diet.

The most common adverse effects of mycotoxins are reduced growth, reduced egg production and increased susceptibility to bruising. In the acute form there may be mortality rates of up to 40–50% of the flock.

Controlling diseases in scavenging flocks

Before it is possible to control disease in rural flocks it is necessary to know what diseases are present and how important these diseases are. It costs money to control disease and it is important to know whether the likely results would justify the inputs of money, labour and time. It is therefore necessary to collect data. These data can be gleaned in a number of ways. The obvious first step would be to interview owners although their memories might not be reliable. This technique would give us some idea of losses that are occurring and possible causes of these losses, as well as information on husbandry practices.

The next step would be to carry out detailed surveys of a number of poultry owned by different owners in different villages. To be of value this type of survey needs to be correctly structured and adequately analysed. One such survey was carried out recently near Gaberone in Botswana. Villagers were asked about flock size, feeding methods and management practices. Two birds were caught (one male and one female) examined for external parasites and fowl pox lesions and scored on a subjective scale of 1–5 (1 = emaciated; 5 = fat with indistinct sternum). 2 cm^3 blood samples were taken from the radial vein by puncture with a 25 gauge needle. Serum was examined for *Salmonella pullorum*, and mycoplasma and Newcastle disease antibodies. In addition to the information gathered in the Botswana survey it is useful to carry out a systematic post-mortem on the carcasses of chickens that have died and to analyse the food that has been fed to the poultry. This type of survey can give a very good idea of the health of birds in the non-commercial sector and therefore what action needs to be taken.

Once the extent of the disease problem has been determined, possible strategies for its control can be devised. The control measures can be at several levels:

- international;
- national;
- village.

Control at international level

It is very easy to import or re-import diseases of poultry into a country. Many developing countries either do not have or fail to enforce regulations on the importation of live chickens, eggs and meat. Countries will wish to import chicks or adults of new improved strains but should be aware of the dangers of importing disease with these birds. It is important that they frame and enforce regulations which reduce the risk of importing diseases and in particular that only disease-free breeding stock is imported.

Control at national level

Scavenger poultry, because they are seldom vaccinated, can act as a monitor of poultry diseases within a country. When an outbreak is confirmed regulatory measures such as the declaration of infected zones, isolation, slaughtering and quarantine should follow. This process is relatively easy to achieve with intensive flocks but much harder to implement with scavenger flocks.

Control at village level

In the village the improvement of the conditions under which chickens are kept can reduce losses due to disease. For example, chicken wire around a compound will prevent chickens straying and keep out salmonella. It may even slow the spread of Newcastle disease as in the tropics the virus lives for a very short time in the environment. Droppings, feathers and dead birds all represent a source of pathogens and parasites which can infect a flock. The removal of these from the birds' enclosure is strongly recommended. Similarly, reducing contact between species can help as ducks, turkeys, geese and guinea fowl can carry diseases which infect chickens. The most certain way of reducing disease is by vaccination. Vaccination of chickens at the village level is often difficult to achieve. The main problem that occurs is the inactivation of the vaccine in the long period between the time when the vaccine is produced and its actual administration to the birds. Vaccines have to be kept cold otherwise they lose their efficacy. The establishment of a system of keeping vaccines cold ('a cold chain') is often difficult to organise. This is particularly true when road conditions and communications are bad as is often the case in developing countries. A solution to this problem in the future may be the development of thermostable vaccines, particularly by some form of genetic engineering. The most important disease in most scavenger flocks is Newcastle disease and as it is endemic in many countries controlling this disease by vaccination would make a large impact on losses by disease.

9 Integration of poultry production into agricultural systems in the tropics

The advantages of scavenger poultry units

In many countries in the world, poultry are kept as scavengers around the houses of the human population (Figs 9.1, 9.2, 9.3 and 9.4). The birds are shut up at night for reasons of security and then released during the day to obtain what food they can from the surrounding land. This food may be supplemented by food left over from the diet of humans although in very poor communities the amount of food available in this form may be very little indeed. In other communities, however, there may be surplus production of cereals, for example in certain parts of the Sudan Zone in West Africa. This system of production, although it appears primitive can be very efficient because although the output from the individual birds is low the inputs are even lower or virtually non-existent. Therefore production from this type of bird is often the only type of production that can be afforded by the poorer members of a society, particularly those living in the rural areas.

The output from scavenging birds is low because of low egg production and low survivability of the chicks. In the south of Nigeria the chickens are raised near the houses and wander into the bush in search of food. No special housing is provided for them. Some families use baskets to take the chickens into the bush during the day and bring them home at night. Some food is given, e.g. in the tropical zone of West Africa termite larvae are given to chicks for the first few weeks of their lives.

Under scavenging conditions the reproduction cycle is comprised of three phases.

(a) Laying phase (ten days).
(b) Incubation phase (twenty-one days).
(c) Brooding phase (about fifty-six days).

Thus the maximum number of clutches is 4.2 per hen each year although 2–3 is nearer the mark in practice. This means that each bird will produce twenty to thirty eggs per annum. Most of these eggs are left to hatch but the majority of the chicks hatched die (Fig 9.5). Consequently the overall production from a flock of 209 hens in Jago village in Nigeria is only about seventy birds per annum (Fig 9.6).

Similar levels of production have been recorded both in the Sudan and

Fig 9.1 *Scavenging poultry*

Fig 9.2 *Scavenging poultry on a Caribbean island*

Fig 9.3 *Scavenger hens and chicks on a government smallholding in Guinea Bissau*

Fig 9.4 *Scavenging poultry on a smallholding in Costa Rica*

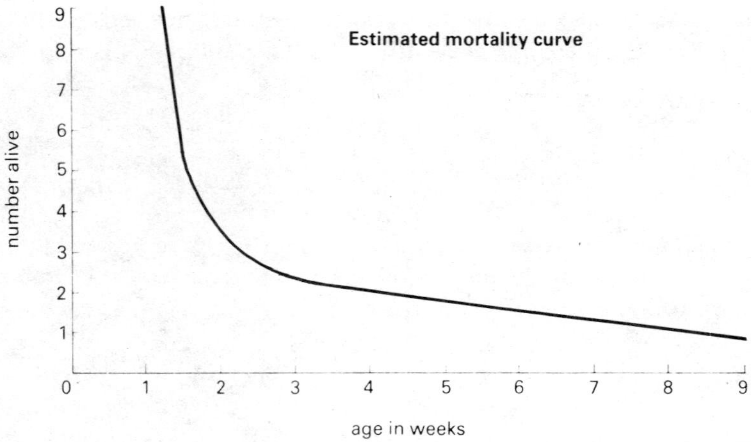

Fig 9.5 *Relationship between age of chicks and number surviving* (data R. Matthewman)

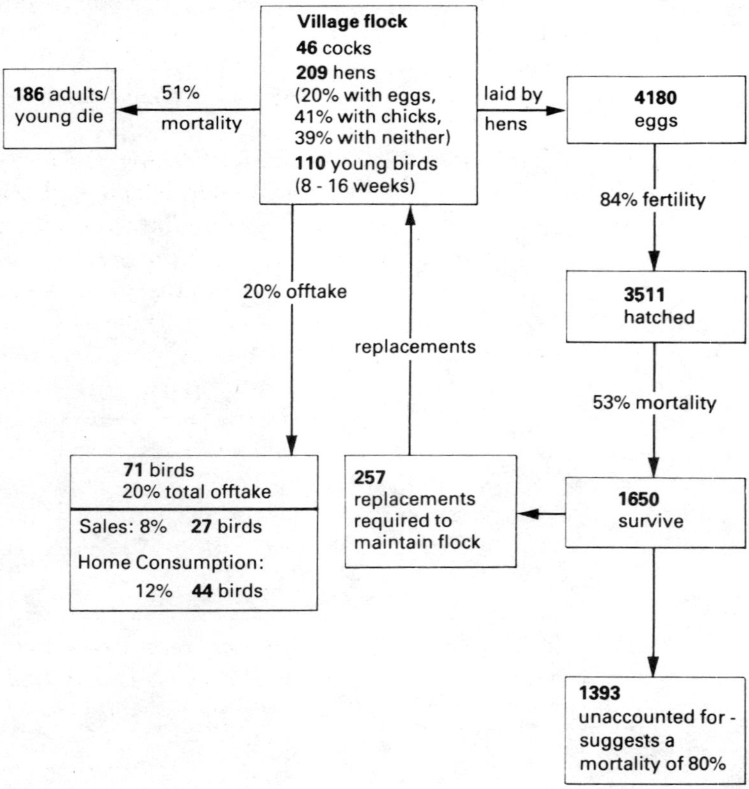

Fig 9.6 *Productivity and offtake of a village poultry flock in Nigeria (Jago village)* (data R. Matthewman)

in Mali. In the former country (Darfur region) thirty-five eggs per bird per year was recorded in village flocks and a similar number in the latter.

Improving the production of scavenging birds

This can be achieved in a number of ways – the use of more productive birds, the provision of supplementary food or better management including disease control and best of all by the combination of all three.

The use of more productive birds

Cockerel exchange schemes have been explored in several countries, e.g. Malawi with the Black Australorps, the Rhode Island Red in parts of West Africa, Pakistan with the Lyallpur Silver Black, and Nigeria with the New Hampshire cockerels.

Under extensive systems of management, improved stock are unlikely to make much impact on the productivity of scavenger birds unless feeding and management are improved. In Bangladesh local desi birds grew as fast as those crossed with White Leghorns, Rhode Island Reds and Fayoumi.

There are of course considerable difficulties in introducing exotic birds into a new environment. White Leghorn and Rhode Island Red hens and cockerels were introduced into villages in Uganda. Their performance was poor and the females ceased laying within a short time under range conditions. White Leghorns were found to be unacceptable to small-scale farmers in Uganda because they were easily seen and killed by kites. Rhode Island Reds and in particular Black Australorps were found to be far more acceptable to the villagers. Any strain or breed that is introduced must have some potential to go 'broody' because farmers will want and need to hatch chicks on their own holdings and will not normally have an incubator for this task. Reducing broodiness by removing chicks from the hens at four weeks of age has been shown in Bangladesh to increase the productivity of desi hens from sixty to ninety eggs per bird per year.

> *Warning: Cases have been reported where villagers have eaten the cockerels soon after they were donated to them, particularly if the birds were in good condition and ready for the pot. If the cockerels are distributed at the time of religious festivals they will be particularly vulnerable.*

If improved birds are to be used it may be better to encourage some form of semi-intensive production. This technique has been used in the eastern hills of Nepal. In a trial carried out in that country, cross-bred and indigenous chickens were kept under semi-intensive conditions. The birds were housed in the local type of poultry houses and each bird was allowed three square metres of scavenging area and a daily ration of 55 g of a commercial food each day. The birds were vaccinated against Newcastle disease and fowl pox. Eggs were removed from the hens soon after laying to reduce the possibility of them becoming 'broody'. The indigenous birds came into lay at the same time (160 days) as the crosses with exotic birds (Light Sussex [LS], Black Australorp [BA] and New Hampshire [NH]) but egg production was considerably lower (125 eggs per annum) compared with that of the LS, BA and NH crosses of 160, 171 and 188 eggs respectively. Under this system of management eleven clutches of eggs were produced each year as compared with three to four which are produced under the normal scavenging system. All the birds kept in the semi-intensive system were fed about 250 MJ a year which provided about a third of their requirements, the rest was provided by their scavenging activities. The cross-bred birds laid large eggs and made a more efficient use of the food fed to them. Thus the increased output under this system occurred partly due to the improved birds but also because of a combination of improved nutrition and better management (removing the eggs from the birds soon after they had been laid).

Improving the diet of scavenging birds

Improving the diet of scavenging birds is difficult because it is not known what food they are scavenging. However, any attempt at supplementation should take into consideration what the birds are actually eating. However even though scavenging birds are often fed either on grain (Fig 9.7) or the left-overs from the diet of humans, which tends to be energy rich, scavenging poultry can have a high protein intake because of their consumption of insects and other invertebrates. That this is so was borne out by the observation of Huchzermeyer (1973). Under subsistence farming conditions in Zimbabwe he found that hybrid poultry kept under free-range conditions produced 170 eggs per bird per year when the diet of the birds was supplemented with maize alone. This result indicated that the birds were obtaining the necessary protein and other dietary sources by their scavenging activities. This conclusion is confirmed by the fact that when the birds were confined for two weeks during the middle of the laying cycle the rate of lay was reduced almost to zero. A supplement of 75 g of maize per egg was needed by the hens to enable them to produce these 170 eggs in a calendar year. Unfortunately Huchzermeyer did not have a control treatment but the rate of production that his birds achieved compares very favourably with the rate of production of the hens in Nigeria recorded by Matthewman (1977), three

Fig 9.7 *Supplementary food for scavenging poultry in Nepal*

clutches of twenty eggs during the year. Part of this improvement was due to better management including disease control and to the use of improved hybrid birds. However the input of 75 g of maize to produce an egg compares very favourably with the input of 130–140 g of a complete diet needed to produce an egg under intensive conditions.

> *There is of course a limit to the number of poultry that can be used to scavenge a given area of land and increasing the number far above this limit would probably lead to a decrease in production per bird.*

Possible new sources of food for scavenging poultry

Small-scale farmers do not normally have surplus food to feed to poultry. Even a small flock of twenty layers kept intensively could consume more than 700 kg of food a year. This is enough food for two adult human beings. Even if only one-third of the food is provided in this way, as is the case in some semi-intensive forms of production, it still represents a considerable drain on the small farmer's resources. The question we must ask now is: are there other sources of food that farmers could use for poultry feeding which would not normally be used for human consumption? Two possible sources of protein are maggots and earthworms, both could be used as a supplement to be fed with foods with a low protein content. These two potential foodstuffs could be used to supplement the diet of birds fed on low quality food such as cassava and sweet potatoes.

Maggots (larvae of *Musca domestica*)

Scavenging poultry almost certainly consume a proportion of their diets in the form of maggots. The Kusasi tribe who live in north-eastern Ghana have developed a technique for growing maggots for poultry feeding. Blood, ruminant contents and dung are mixed in a large open pan and this pan is placed above a collecting container (Fig 9.8). The smell of the mixture attracts flies which deposit their eggs. The maggots are well nourished by this mixture. When they are ready to pupate they migrate out of the dung mixture and fall out of the pan into the collecting containers. The farmers then collect the maggots, wash them and feed them to the birds. As the feeding value of the mixture declines the number of maggots harvested per day also declines.

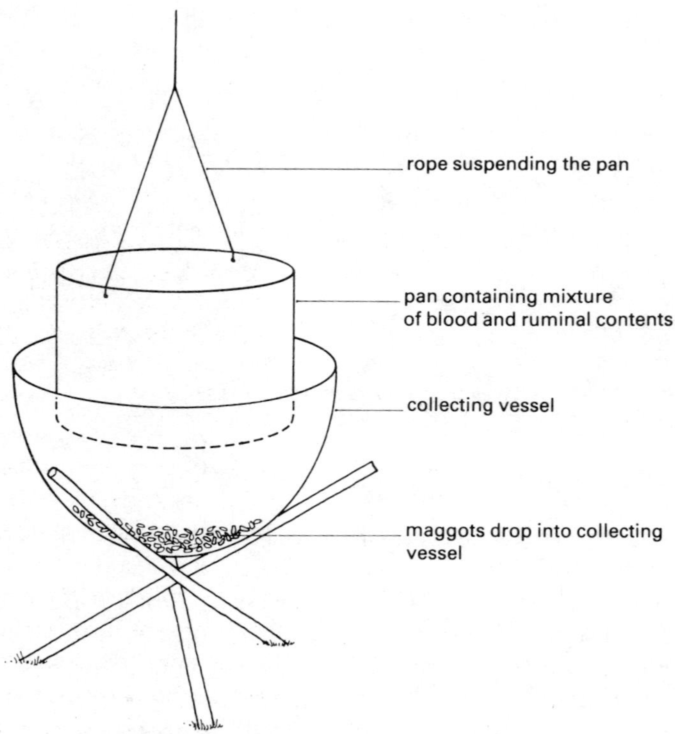

Fig 9.8 *A system for growing and collecting maggots* (after B. Korugyendo)

Maggots have a high content of crude protein. Their dry matter which represents 30% of the total mass of the larvae contains 54% crude protein. Assuming that the maggots are fed alive and that they constitute the main source of protein, each laying bird will need to eat approximately 100 grams of maggots each day. However, because birds will consume protein from other sources, 50 grams a day of live maggots should be more than enough to meet the needs of laying birds. However, even this comparatively small intake would mean that farmers would have to supply 360 kg of maggots each year to meet the needs of a flock of twenty laying birds.

Earthworms (*Eisenia foetida*)
Earthworms have a reasonably high protein content (55% of the dry matter) but also have a high water content (84%). Earthworms grow well throughout the world and are favoured by a hot humid environment. The faeces of cattle are a good medium for growing earthworms. It has been claimed that the dung from one high-producing cow will produce earthworms with a total crude protein content of 425 g. This would be enough protein for fifty laying birds, if the protein was used to supplement other foodstuffs which contained some protein. Experiments have been carried out which indicate that earthworms are a suitable foodstuff for feeding to poultry as the sole or main protein supplement.

Changes in management techniques

The productivity of scavenging birds can also be improved by relatively simple changes in management technique which will either promote productivity, reduce mortality or increase the number of eggs harvested. For example if laying birds are kept in their houses until 10.00 h and fed whilst in the house, most of the eggs produced will be laid within the nest boxes in the house and not in the surrounding bushes. This will happen because most eggs are laid before 10.00 h. Almost all of these eggs should be removed from the nest boxes to reduce breakages and also to encourage the birds to lay more eggs. This management technique will considerably increase the output of eggs. A few eggs should be left in the nest to encourage the birds to lay in the nest box provided in the house (Fig 9.9).

Clean water should be provided at all times. All birds, and particularly chicks, can quickly dehydrate. Chicks should be protected during their initial stages as most die during the first two weeks after hatching. If they can be kept alive at this stage a considerable increase in the output of chicks will be achieved. A simple chicken run can be used to protect the chicks from rain and predators (Fig 9.10).

Disease and parasites should be kept under control. If possible routine

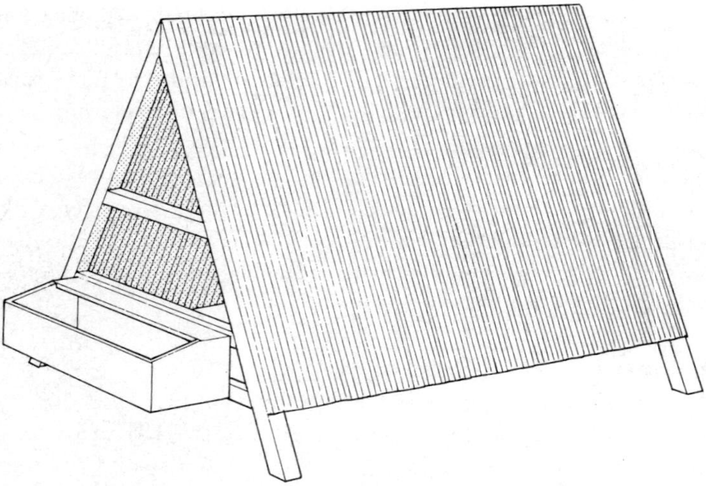

Fig 9.9 *A simple field ark for laying birds* (Chris McKinnell)

Fig 9.10 *Box at the end of a small chicken run to protect the chicks from rain and predators – the wire netting can be replaced with reeds, elephant grass or papyrus reed*

vaccination against Newcastle disease and fowl pox should be given and prophylaxis against coccidiosis and helminths.

Keeping birds in fenced runs is not a good idea (Fig 9.11). The farmer has to provide all the food that the bird needs since the birds do not have the opportunity to scavenge their own food.

If the housing is not easy to keep clean (Fig 9.12) external parasites may become a problem. Inadequate nesting facilities (Fig 9.12) may encourage cannibalism leading to prolapse of the reproductive tract.

Fig 9.11 *Semi-intensive farming in Zimbabwe (left)*

Fig 9.12 *Good ventilation, but hard to keep clean (note inadequate 'nest') (right)*

Scavenging geese

A pair of geese with access to good quality grazing and water can produce 45–75 kg of meat for a family each year for a period of twenty years or more. They can be kept in crops such as coffee, bananas, pineapples and other cultivated crops which grow tall enough to avoid the problem of the geese consuming the crop along with the grass. Provided they are given adequate food, ganders should weigh up to 9 kg at nine months of age and geese up to 7.5 kg. If this food is provided from grass, five geese will consume as much grass as one sheep. They can also be fed intensively on diets of the type fed to ducks, in which case they will reach 6 kg in twelve weeks with a conversion ratio of about one kilogram of liveweight for every kilogram of food consumed.

Breeding geese

Geese are not normally mated until they are six to eight months of age. One gander is kept with three to five geese. The fertility of geese is influenced by day-length and they normally produce eggs under long day or increasing day-length. The eggs produced should be collected each day and incubated artificially since geese are not particularly good mothers.

The egg takes twenty-eight to thirty days to hatch. Incubation can be carried out either under broody hens or in incubators. If hens are used the eggs should be turned once or twice a day since hens are unable to do this themselves because the eggs are too large. The temperature in the incubator should be lower than that used for hens' eggs (30°C). Relative humidity should be in the region of 80%.

Rearing goslings

The rearing of goslings is similar to the rearing of ducks except that since they are larger, fewer can be kept in each brooder. As with ducks artificial heating can be removed after the goslings are two weeks old, they should however be provided with shelter.

A feeding regime based on that used for ducklings is satisfactory but they can start to make use of grass at two to three weeks of age. It should be remembered that geese are naturally nocturnal feeders, so where the absence of predators permits the birds should be allowed to graze at night. Water for swimming is not required but geese of all ages should be able to immerse their heads in water when they so desire. Geese of most breeds are ready for market by fifteen weeks of age. Goose meat contains considerably more fat than meat from other poultry, but is well accepted by local populations including some of those in West Africa. The higher fat content cannot be completely overcome but it can be reduced by using Chinese geese which contain less fat than other breeds.

Geese, particularly of the lighter breeds, can fly therefore the feathers on one wing should be clipped.

Scavenging ducks

Scavenging ducks are particularly important in south-east Asia. Throughout China, Thailand, Malaysia, the Philippines and Indonesia ducks are kept to provide eggs and meat. Ducks can forage for food and as a consequence the system of driving ducks through fields exists in all south-east Asian countries.

Ducks like water and eat small animals or snails and therefore other systems of keeping ducks near or even on water are employed. The ducks are allowed to forage during the daytime and are normally penned to lay eggs at night.

On the island of Java, Indonesia, ducks are allowed to scavenge on paddy fields and consume fallen rice grains and green material that grows on the fertilized and irrigated rice fields. They are allowed to scavenge during the day but are penned from late afternoon to early morning. Eggs that are laid during the period when the ducks are penned can be collected. Their daily cycle is repeated until all the food that can be scavenged is exhausted and they are then removed to a new site. Experimental work in Indonesia has shown that rice grains account for 75% of the crop fill of these ducks and aquatic molluscs for about 20%. The crude protein content of the dried crop content has been estimated as being 10–12%. The scavenging system is sometimes intensified in Thailand and the Philippines. In these countries ducks are sometimes totally confined and fed on low cost by-products such as snails, dried or smoked fish, coconut meal and sago palm. Sometimes this system is combined with fish farming.

In China, ducks are used to reduce crop damage caused by pest infestation, to remove weeds from the banks of waterways and to clear aquatic weeds out of ponds and canals. However, ducks will also eat crops and may become pests causing considerable damage to cereal crops.

The integration of poultry with fish farming

In developing countries large numbers of farmers have only very small farms to cultivate. Thus the maximum use of limited resources is of great importance as is the proper integration of several enterprises. For the successful integration of poultry with other enterprises such as fish and crop production, a compatible arrangement must be made. Examples of compatible enterprises including poultry are given below.

- Swine – chicken, fish
- Duck – fish
- Poultry – fish, rice, forage, other crops

In swine/chicken/fish systems, the animal manure is used directly to feed the fish. Handling the animal waste is made easy and the waste is turned into a useful commodity.

In integrated duck/fish farming the ducks make use of the pond water surface and adequate food in the shallow parts of the pond whilst the fish thrive on the droppings of ducks which fertilize the ponds. About 25 kg of duck manure will be needed for each kilogram of fish flesh. One duck produces about 72 kg of droppings in a year. Therefore 500 ducks kept in a one hectare pond will produce 36 tonnes of manure in a year sufficient to yield 1400 kg of fish.

Another version of the integration of ducks with other forms of production is seen in eastern Europe. In some of these countries, ducks and fish are farmed together for five to six years. After this period a build up of organic material leads to reduced fish production. At this point the fish ponds are drained and crops such as rice are grown on the empty pond beds for two to three years after which time the duck/fish farming system is recommenced.

10 Marketing poultry products

Introduction

When a farmer produces eggs and/or poultry meat over and above the requirements of his family these products become a liability unless a market can be found for them. Most farmers think primarily in terms of production and only secondly of the means of disposing of surplus production. The problem of this surplus production will vary greatly between countries and even between regions within countries. Also there is a great difference between the marketing of poultry meat and the marketing of eggs produced by poultry. In many countries of the developing world where there is a shortage of animal protein and where the farmer is located near a large centre of consumption then disposal of poultry meat is not a problem. Consumers will often queue at the door of the producer to take away the product that has been produced that day. The only problem arising then is one of avoiding theft and setting the market price at a level which the consumer will be prepared to pay and which will maximise income. On the other hand if the farmer is in a rural area away from a main market then the cost of transporting his products to a market may be exorbitant and render the selling of surplus produce uneconomic.

Marketing eggs

Eggs are a very valuable source of animal protein and a very convenient one, easily packaged, easily distributed and needing no preparation before consumption, apart from cooking the product in one way or another.

Because eggs are such a good foodstuff they are also a good food for bacteria and therefore can easily become infected and lose their quality or become positively dangerous if incorrectly handled after laying. It is therefore important that eggs should be produced in as clean a form as possible so as to avoid spoilage and to avoid the need for washing which can remove the protective coat from the outside of the egg. However, normally some eggs will require to be washed and if this is necessary, the following process is advised.

Cleaning of eggs

Only eggs that are dirty should be cleaned and cleaning them with a damp rag should be avoided at all costs as this is likely to lead to them becoming contaminated even more quickly than if they are left dirty. However, some form of cleaning is essential if they are to be marketed. Eggs covered with poultry manure or the remains of broken eggs are not easily sold except in times of extreme shortage. The best way of cleaning eggs is by immersion cleaning in water containing a sanitiser. The temperature of the water containing the sanitiser must be higher than that of the eggs and this means that the temperature should be between 33°C and 35°C in temperate climates but 40°C in the tropics. At the lower temperature an immersion time of ten to fifteen minutes is necessary to thoroughly sanitise the eggs but at the higher temperature the immersion should be limited to three to five minutes particularly in the case of eggs that are to be retained for hatching because at the higher temperature the embryo may become damaged by longer periods of immersion.

The eggs can be immersed in the water containing a sanitiser in a plastic mesh bucket and rotated gently backwards and forwards. They will then become very clean in the time allocated. This is obviously most efficiently done by a mechanical egg cleaner but in the absence of one of these it would be relatively simple to design a hand-driven machine which would achieve the same result. The amount of sanitiser put into the water should be that advised by the manufacturers and a number of different materials, with different trade names, are suitable. Compounds based on the quarternary ammonium compounds, chlorine-based compounds, or iodophor compounds should be used. Water used for cleaning should be changed frequently to minimise the risk of contamination.

Storing the eggs

In the tropics eggs can very quickly deteriorate unless they are stored at a low temperature. Eggs that are being stored for incubation will start to incubate unless they are kept in a cool house. The ideal temperature for storage is lower than 13°C but this is often impossible in tropical countries. In these countries the eggs should be kept out of direct sunlight and left in the coolest place possible and sold by the third day after laying. In the absence of mechanical refrigeration, putting the eggs in a porous pot where the outside is kept damp is a very useful cooling mechanism but care should be taken to avoid the excess use of water and making the bottom eggs in the pot too wet. Eggs absorb strong odours very quickly and should not therefore be stored in the food room where fish-meal is being kept nor should they be kept stored in a freshly painted room where they would absorb the smell of the paint.

Egg grading

In sophisticated markets eggs are often graded by size and by shell shape. Only those eggs which have smooth shells and with a normal egg shape are sent to market. In some such markets brown eggs are preferred to white ones.

In most countries where eggs are in short supply, grading is not necessary and should only be carried out if it can guarantee an increased income because grading obviously takes time and costs money.

In some countries, populations will not consume fertilised eggs. This is true with some Hindu peoples, whereas in other countries, notably in African countries, only fertilised eggs are consumed. These preferences should be taken into consideration when eggs are being prepared for sale.

The production of eggs in intensive units is normally more expensive than their production from scavenger units but in many countries there is a seasonal supply of eggs from these latter units. When these eggs come onto the market it is found that the price of all eggs tends to decline. Therefore the producer of the intensively produced eggs should ensure that his maximum supply coincides with the minimum supply of eggs from the scavenger units. If he does this he will be likely to maximise his income.

It should also be remembered that it is very easy for the number of eggs being produced within a country to be increased very quickly indeed. One bird of the modern hybrid type can produce 280 eggs in a year and with the levels of consumption that are presently occurring in many developing countries, this supply of eggs would be sufficient to meet the demands of five people. With rapid expansion it is very easy to go from a deficit to a surplus supply in one or two years and if large amounts of capital are put into poultry production it would be very easy for farmers that are producing these eggs to go bankrupt. Therefore, a careful assessment of the likely market, the changing market and changes in other farmers' levels of production should be made before large amounts of capital are spent on developing poultry units.

Marketing poultry meat

As has been noted before, over the past fifty years, there has been a marked change in developed countries from dual-purpose poultry that produce both meat and eggs to single-purpose poultry that produce either broiler meat or eggs for human consumption. This change has not been so complete in most developing countries except where intensive poultry production has been established around large towns. In many countries meat is still produced from dual-purpose local birds which may be one-and-a-half to two years of age, before they are slaughtered. Meat produced from this type of bird tends to be stringy but has plenty of 'chewability'. The traditional way of preparing such a bird for consumption is by boiling in a pot. If such a process is attempted with a modern broiler the meat will disintegrate and soup rather than cooked meat will be produced. Therefore, if it is intended to introduce this new type of bird into an area where it has not been used before, not only is marketing required but also consumer education on the preparation of the meat is highly desirable.

The farmer also has to decide whether or not he will market the meat as live birds or whether he will slaughter and prepare the birds for the table. This decision will depend on many factors and especially on the sophistication of the markets. Other factors which have to be taken into consideration are the religious beliefs of the people who are buying the birds. For example, factory killed birds may not be acceptable to consumers of the Moslem faith. Also once the bird has been killed the carcass will deteriorate rapidly unless it is subjected to refrigeration whereas the live bird can be transported and kept for an indefinite period of time provided it is fed.

Producers should also be aware of market trends for poultry meat. It may be, for example, that poultry meat is only eaten in large quantities on certain religious festivals or possible only at cooler times of the year and therefore production should be geared to coincide with these demands. It is however, possible to change demands for meat. For example, in Britain the consumption of turkey meat used to be a very seasonal one coinciding with the month of December when the Christmas holiday occurs. However, with the reduction in the price of turkey meat and the advent of sophisticated marketing techniques, although the bulk of turkeys are still consumed at this time of the year, there is now a much more even demand for the birds throughout the year. Such an even flow makes it much easier to organise production and avoids having equipment left idle for long periods of the year when birds are not required or alternatively it avoids the cost of storing these birds in a deep freeze. This spread of demand has been achieved partially by selling parts of turkeys as turkey joints rather than whole turkeys. Jointing of the turkey is carried out because it is recognised that families in Britain are small and that a turkey is too large for one family. Therefore parts of a carcass are preferable to the whole bird just as beef is sold in joints rather than as a whole carcass. Such a ploy is less likely to be of value in developing countries where families are large. Of course incomes are often low and therefore a small part of a bird may be desirable in order for the purchaser to be able to afford to buy any poultry at all.

The marketing of either eggs or poultry is obviously a very time consuming operation and it may be that the farmer does not wish to involve himself in such an operation. It may be possible to reduce the amount of time that he would have to spend on these operations by handing over the marketing side to a co-operative which can organise the grading of eggs and the slaughter of poultry and also ensure that the grading of the product is of an adequate standard. Such a co-operative can also negotiate large-scale contracts and purchase refrigeration equipment to store produce during the seasons when the products are fetching a low price. They can also provide advice to the small-scale poultry keepers in the locality.

Preparation of poultry carcasses for consumption and sale

Catching and transporting

All birds destined to be slaughtered should be fasted for the previous four to six hours to allow the intestine to be emptied. A longer period than this should be avoided otherwise unacceptable weight loss will occur. If poultry are not starved before slaughter the gut may rupture when the carcass is eviscerated.

The bird should be caught under the least stressful conditions possible. In the case of windowless houses this may be achieved by catching the birds under a blue light if they have previously been reared under white light or red light. Under tropical or subtropical conditions the best solution may be to catch the birds at dusk or at night. When catching small numbers of birds the use of a specially made crook can be invaluable (Fig 10.1).

The birds should be placed into specially made crates where possible. They are put in head first in the manner shown (Fig 10.2). These crates should have dimensions of 80 cm × 60 cm × 30 cm. Crates of these dimensions will hold ten mature broilers or adult laying birds. Once the birds have been caught and put into the crates they should be kept out of the sun and in a well ventilated environment, and not transported during the heat of the day. Long journey times to the processing factory should be avoided.

Killing poultry

On the small farm poultry are killed by breaking the neck. This is achieved by holding the chicken by its feet and pulling downwards on the neck and bending the head sharply backwards. A considerable amount of force is needed to break the neck and to leave a space into which the blood can drain. The ability to kill poultry by this method, once learned,

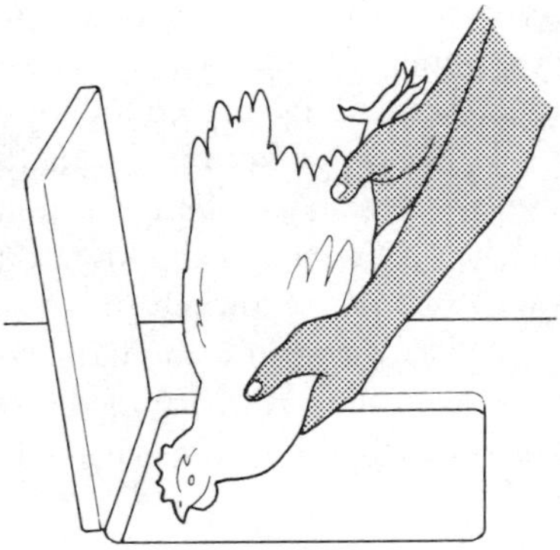

Fig 10.1 *Use of a crook to catch mature birds (left). The bird is approached from behind and the crook is placed just above the foot – on reaching forward the bird can be picked up gently.* (Courtesy of *ATB*)

Fig 10.2 *Placing a bird into a crate (right). The bird should be lowered in gently, head first, holding the wings close to the body and the legs above the joint.* (Courtesy of *ATB*)

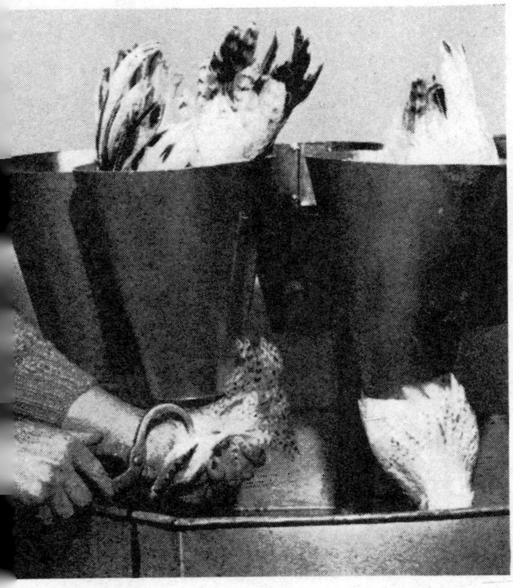

Fig 10.3 *Chicken in a bleeding cone*

makes the killing process simple and quick. If too much force is applied the neck will be severed and blood will be sprayed around the slaughterhouse. If poultry are killed for use by Muslims the head should be removed with a single cut. Another approach to the slaughter of poultry is to immobilise the chicken in a bleeding cone (Fig 10.3) and to cut the jugular vein to ensure complete exsanguination. The cut is made either below the ear or within the mouth.

In an industrial plant the poultry are normally stunned by passing their heads through a shallow trough of electrified water. This is achieved as part of a conveyor belt process in which the birds are moved around a factory on a continuous chain whilst various carcass preparation operations are carried out. Immediately after stunning the birds are killed by cutting the jugular vein with a knife.

Removing the feathers from poultry

On the small farm and under traditional farming conditions, chickens will normally be plucked by hand and one person should be able to pluck four to ten chickens per hour. Dry plucking as the above is called is a fairly slow, dusty operation. The feathers can be removed most easily when the body is warm. The large wing feathers are removed first, next the back, sides and the abdomen, the breast feathers must then be removed very carefully to avoid tearing the skin. Lastly the legs, neck and wings are plucked and the abdomen squeezed in order to remove any droppings in the vent. This will help to prevent 'greening' of the flesh and the danger of taint. The process can be speeded up if the birds are immersed in hot water after killing. The temperature of the hot water should be 52°C for broilers, 55°C for old laying hens and 59°C for turkeys. They should be immersed in the water for approximately two minutes.

The operation can be further speeded up by the use of mechanical pluckers. With these machines the feathers are removed by contra-rotating cylinders on which supple rubber figures are mounted. Such machines may be purchased by co-operatives and by farmers producing birds on a large scale.

Carcass preparation

Once the feathers have been removed the intestines should be removed. This is achieved by drawing the intestine out through the cloaca. Sometimes the oesophagus, crop, gizzard, heart, liver and lungs will remain in the carcass, but in the factory operation most of these items are removed and the abdominal contents thoroughly washed out before the rest of the operations are carried out. In this operation the feet will also be removed and the wings and feet tied down to leave a neat carcass for easy marketing.

On the farm the operation is carried out as follows, first the leg sinews are drawn out from the legs and the legs are cut off below the knee joint. The neck is then severed close to the body leaving a flap of skin to close the hole. Through this hole the crop is removed and at the same time the lungs are loosened so that they can be removed through the rear of the carcass. The intestines can now be cut free and the remainder of the body contents removed. To make a tidy carcass the legs can be tied close to the body by passing string through the body with the help of a needle below the end of the breast bone, around the legs and around the tail.

Cooling carcasses

Most of the preparation of the carcasses is carried out at normal body temperatures. However, as soon as this process is completed it is important that the carcasses be cooled as soon as possible to limit the extent of bacterial growth. The carcasses can be cooled by one of two methods.

- Putting the carcass in iced water for a minimum of thirty minutes to reduce the temperature of the carcasses to not more than +4°C. This is not a particularly hygienic process since bacteria can easily be transferred from one carcass to another – outbreaks of salmonellosis (food poisoning) in humans can be caused in this way.
- A more satisfactory method of chilling the carcasses is in a refrigeration plant, if possible one that uses a forced draught. The incoming air needs to have its moisture content closely controlled to avoid excessive moisture loss from the carcasses.

Storage

The temperature at which the carcasses are stored is dependent to a degree on the length of time for which the carcasses are to be stored. If they are to be consumed within a day or two they can be stored in a refrigerator at +2°C. If they are to be stored for a long period of time they should be frozen to −5°C or even to −18°C.

In many parts of the world freezing birds is not a practical proposition nor is it necessary. The birds are kept alive until they are needed for consumption and then they are killed. The amount of meat on a carcass is so limited that it can easily be consumed by an average family in one meal.

11 Maximising profits from poultry

Introduction

If poultry are kept in the form of a scavenger flock, normally the inputs are regarded as zero because food is not purchased and the birds are fed on by-products from the household which probably do not have an alternative use. However, once a farmer starts to produce either poultry meat or eggs for sale then, of course, inputs are purchased and the farmer becomes interested in obtaining enough income to cover the cost of these inputs, plus the cost of depreciation on capital goods purchased, plus a return for his labour. Under these circumstances it is important that the profits achieved are as high as possible. The most important ways in which profits are maximised are by ensuring the following.

The efficient use of food

Food is the most important input into a poultry enterprise representing between 60% and 70% of the total costs of production in many poultry flocks. Total costs of food can be reduced by ensuring that the cheapest sources of food ingredients are found, provided this does not result in lower efficiency of use due to the utilisation of inferior ingredients. Also it is very important that food is not wasted. The design of feeders has been discussed previously but it is important to encourage the operatives not to fill troughs more than one-third full otherwise large quantities of food will be spilt by the birds when they eat. Food is also a very saleable commodity and theft could be a very serious problem. It is important, therefore, to ensure that adequate records are kept and that security of the food store is maintained. If this is not done the profitability of the enterprise will be jeopardised.

Maximising production at the correct time of year

In many countries the price of eggs and/or poultry meat may not be the same throughout the whole year. For example, if there is an influx of eggs from scavenger poultry at certain times of the year the presence of these eggs will depress the price of eggs produced from intensive units. It is therefore desirable for the intensive producers to ensure that most of the eggs are produced by their flocks at times of the year when the production of eggs from the scavenger flock is at a minimum. This can be achieved by the process of buying chicks at the right time of the year and also by systems of lighting the birds. It is also important to ensure that the birds produce as many eggs or grow as fast as possible to minimise the amount of food that is used for maintenance. For example a 2 kg laying hen will consume about 30 kg of food each year just to maintain its body weight. This food is a tax on the system and the bird will need about another 16 kg of food to produce 250 eggs in a year. Any rationing of food will merely ensure that the bird decreases production and therefore rationing or any form of restriction is going to lower profitability.

Using good quality labour

It is almost always a false economy to use cheap under-trained labour. If the labour is responsible for the management of the flock and the labour has limited knowledge of flock management this can very quickly render the flock unproductive and therefore unprofitable. Labour should be trained in proper feeding techniques, in lighting, especially for laying birds, and in the management of the birds to encourage the removal of diseased birds and birds that are not laying. A non-laying bird in a laying house is a tax on that house.

Using the right type of bird

It is normally a false economy to use cheap birds for intensive methods of production. Specially bred hybrid birds when used for either laying or for meat production, although they cost more, are usually far more productive and make better use of their food and therefore use the resources more efficiently. The criteria that should be used for the provision of scavenger birds are somewhat different, in this case hybrid birds would not be used and some form of dual purpose bird is to be preferred. In both cases the purchase of cheap birds from unreliable suppliers is to be discouraged. Such birds can easily introduce diseases into the flock.

Accounting for inputs, outputs and profits

Before starting a small or large poultry enterprise it is advisable to draw up a business plan. Such a plan, giving details of the expenses to be incurred of a capital and variable nature both before and after money starts to come into the enterprise, is essential if money is to be borrowed from a bank or government. No sensible lender of money will lend money unless he is reasonably sure that the enterprise is sound and well planned. Details of drawing up a business plan can be obtained from a government advisory officer or from personnel in firms selling food or chickens to the farmer.

The actual accounts

The type of costs that have to be taken into account include fixed costs (cost of buildings and equipment and repayment of loans) and variable costs (cost of birds, food, repairs, labour, fuel and veterinary supplies). The costs have to be covered by the sale of birds and/or eggs. The difference between the money received from sales and that spent on the variable costs represents the **gross margin** received on that particular batch of birds. The **net profit** is the difference between the income from sales and the sum of the variable and fixed costs. In developing countries where buildings are simple and other capital equipment is kept to a minimum the overhead charges should be low. It is normally inadvisable to spent too much money on fixed items, particularly if the money has to be borrowed and the market prices are uncertain. If the overheads are kept low the principal cost of a poultry enterprise is the cost of food which may represent 60% of the total cost of production. The ratio between the cost of food and the cost of the finished product is likely to be the major factor affecting productivity. The price of the chick both for broiler and laying hen enterprises is also likely to be important especially where improved strains are used. These may represent up to 15% of total costs. This is important in the case of broilers but less important in the case of laying birds, because in developing countries the laying bird at the end of the laying period can normally be sold, even if it is very fat, for more than the price of the chick which replaces it and even exceeding the total cost incurred in rearing pullets to point of lay.

As well as minimising costs and maximising output, the farmer should market the product efficiently, when prices are highest.

Records

It is not possible to assess either efficiency or productivity unless adequate records are kept. The records should not be too elaborate or else they may not be kept properly by the workers. Problems may be experienced where workers are illiterate. The type of records that can be kept both for management purposes and to assess profitability are given in Tables 11.1, 11.2, 11.3, 11.4 and 11.5.

Table 11.1 Expenses involved in the rearing of laying hens to point of lay (19 weeks)

Laying hens			
Chick mash	tonnes × price of food	=	
Rearing diet	tonnes × price of food	=	
Fuel costs for the batch of birds		=	
Cost of chicks		=	
Cost of vaccination and medication of the batch of birds		=	
Labour costs		=	
Total variable costs of producing point of laying birds		=	________
Housing (depreciation over 10 years) (allow for number of times house is used each year)		=	
Equipment (depreciation over 6 years) (allow for number of times house is used each year)		=	
Interest on loans		=	
Total fixed costs		=	________
	Total costs	=	________

Table 11.2 Expenses and returns involved in operating a laying house

Expenses			
Cost of point of lay pullets (carried forward)		=	
Food for laying period	tonnes × price of food	=	
Fuel		=	
Transport		=	
Labour		=	
Vaccination, medication		=	
Total variable costs of rearing and laying periods		=	________
Housing (depreciation over 10 years)		=	
Equipment (depreciation over 6 years)		=	
Interest on loans		=	
Total fixed costs		=	________
	Total costs	=	________
Income			
Total sale of eggs		=	
Sale of manure		=	
Sale of culled hens		=	
	Total income	=	________
	Profit	=	________

Table 11.3 Expenses and returns in producing broilers or other poultry for consumption

Expenses			
Cost of chicks		=	
Starter mash	tonnes × price of food	=	
Finisher mash	tonnes × price of food	=	
Fuel		=	
Labour		=	
Vaccination, medication		=	
Collection, transport and slaughter costs		=	
Total variable costs		=	______
Housing (depreciation over 10 years)		=	
Equipment (depreciation over 6 years)		=	
Interest on loans		=	
Total fixed costs		=	______
	Total costs	=	______
Returns			
Sale of broilers		=	
Sale of manure		=	
	Total income	=	______
	Profit	=	______

Table 11.4 Record card for rearing or broiler chickens

Name of operative: ______ Number of house: ______

Date of stocking house: ______ Date of destocking: ______

Number of birds and type: ______

Date of vaccination: Newcastle ______

Marek's ______

Fowl pox ______

Other ______

Dates and causes of death: ______

Total deaths and culls: ______

Amount of food delivered and dates: ______

Total food consumed: ______

Fuel used and dates: ______

Number and total weight of birds removed: ______

Table 11.5 Laying house record card

Name of operative:	________	Number of house:	________
Date of stocking house:	________	Date of destocking:	________
Number of birds and type:	________		
Dates of vaccination:	Newcastle	________	
	Marek's	________	
	Fowl Pox	________	
	Other	________	
Dates and causes of death:	________		
Total deaths and culls:	________		
Amount of food delivered and dates:	________		
Total food consumed:	________		
Number and total weight of birds removed:	________		
Egg Production:	January	________	
	February	________	
	March	________	
	April	________	
	May	________	
	June	________	
	July	________	
	August	________	
	September	________	
	October	________	
	November	________	
	December	________	
	Total	________	

NB This card should be used in conjunction with cards recording daily production of eggs

Glossary

Absolute humidity (Also called vapour concentration; vapour density.) The ratio of the mass of water vapour present in the air to the volume occupied by the mixture.
Acclimation The adaptive changes in response to a single climate variable (normally produced in climatic chambers).
Acclimatization A long term physiological adjustment which results in an increased tolerance to continuous or repeated exposure to complex climatic stressors (normally produced under field conditions).
Aestivation A state of torpidity induced in some animals by the heat and dryness of the summer.
Allele Different forms of a gene that occupy a particular locus (place) on a chromosome.
Antibiotics Substances that kill bacteria. They have been produced by certain other micro-organisms, the best known of which is penicillin.
Apparent digestibility The amount of dry matter of a nutrient ingested minus the amount excreted in the faeces, whether a residue from the food or of endogenous origin, expressed as a percentage of the intake.
Asphyxia Lack of oxygen in the blood caused by restricted respiration.
Assortative mating Mating like to like, i.e. mating the animals according to their phenotypic similarity.
Available protein The amount of truly digestible crude protein in the food less the amount of that protein required to make good the loss of nitrogen as metabolic faecal nitrogen that the ingestion of the food necessarily entails.
Bacteria Microscopic, unicellular organisms that reproduce asexually by fission.
Biological value (BV) The percentage of the truly digestible nitrogen of the food protein that is retained by the animal and used for maintenance, growth, production and pregnancy.
B.Th.U. The British thermal unit of energy, which is the heat required to raise the temperature of one pound of water one degree Fahrenheit (252.1 calories or .1055 joules).
Calorie A unit of heat equal to the amount of heat required to raise the temperature of 1 g of water through 1° Celsius (approx. 4.19 joules).

Character A term used, often rather vaguely to designate any form, feature or function of an organism.
Chromosomes The carriers of inheritance located within the nuclei of cells within the living body.
Closed flock A flock into which there has been no introduction of outside blood since its foundation.
Coccidiostat A drug which when fed in the food will help to prevent the development of an outbreak of coccidiosis in poultry.
Conversion *Celsius to Fahrenheit*: add 40; x $\frac{9}{5}$; subtract 40. *Fahrenheit to Celcius*: add 40; x $\frac{5}{9}$; subtract 40.
Correlation of traits Some characters are genetically correlated. This correlation can be positive or negative, e.g. rapid gain in poultry is **negatively** correlated with egg production. (Negative correlations of favourable characters make improvement by selection more difficult.)
Criss-cross breeding The alternate use of sires of two different breeds generation by generation.
Crude protein All proteins contain about 16% nitrogen. Thus the amount of nitrogen in a food multiplied by 6.25 gives the crude protein.
Cyanosis Bluish-purple discolouration of skin and mucous membrane arising from a deficiency of oxygen in the blood.
Dead vaccine A vaccine containing dead cells of pathogens which can induce immunity into a bird (or other animal).
Dew point (dew point temperature) The temperature to which a given parcel of air must be cooled at constant pressure and constant water vapour content in order for saturation to occur.
Diarrhoea Excessive secretion of water in the faeces.
Digestible protein The part of the crude protein which can be digested and used by the animal.
Diurnal Daily.
Dominance In an individual which is heterozygote for a dominant gene, the phenotype is indistinguishable from the homozygote.
Energy retention The energy retention is the heat of combustion of the food minus the heat of combustion of the faeces.
Epistasis Interaction between genes at different loci on the same chromosome.
F_1 The first generation produced by mating parent stocks.
F_2 The second generation, produced by crossing males and females from the F_1 generation.
Family selection Selection of individuals on the performance of their relatives (sibs, half-sibs, or progeny). Hence selection between families instead of between individuals.
Foot candle 1 foot candle = 1 lumen per square foot = 10.76 lux.
Gene A unit of hereditary transmission located on a chromosome.
Generation interval or generation length The average age of an

animal at the birth of its progeny.

Genetic correlation That part of the phenotypic correlation between two characters due to genetic causes.

Genotype The genetic constitution of an individual with respect to any limited number of genes under examination.

Genotype/environment interaction Situation in which the magnitude of genetic differences for any trait is influenced by the environment where the measurements are made, e.g. breeds of poultry that produce well under intensive management may not be genetically superior to other breeds under village conditions.

Half sibs Half-brothers and -sisters. The offspring of a single sire out of different dams (or vice versa).

Heat increment The increase in heat production which results when an increase occurs in the intake of metabolizable energy.

Heritability *Narrow (and usual) sense* – The fraction of the observed superiority or inferiority of selected animals which reappears in their offspring. *Wider sense* – The ratio of genetic variance to phenotypic variance.

Heterosis The extent to which the progeny of a cross differ from the average of the parental types in a given character. Positive or desirable heterosis is called 'hybrid vigour'.

Homozygote Individual derived from the union of identical genes at one or more loci.

Hybrid a) Sterile offspring of genetically incompatible parents of different species, e.g. male ducks produced by crossing ducks with Muscovy ducks. b) Heterozygous offspring, also termed cross-breds. c) Commercial strains produced by mating different inbred lines to produce heterosis.

Hybrid vigour Positive heterosis.

Hybridisation Mating of animals of different species.

Hyperthermia The state when the body temperature rises above normal.

Hypothermia The state when the body temperature falls below normal.

Imperial/metric equivalents.

1 acre	– 0.405 hectares
1 pound	– 0.454 kg
1 lb/acre	– 1.121 kg/ha
1 ton/acre	– 2.513 tons/ha
1 gallon	– 4.546 litres

Inbred line Line obtained by continued close inbreeding and consisting of individuals presumed to be more or less homozygous for the same genes.

Inbreeding Mating of individuals more closely related than the average relationship of the population to which they belong.

Inbreeding depression Decline in vigour and associated characters due to inbreeding.
Incross Progeny of mating between inbred lines of the same breed.
Incross-bred Progeny of mating between inbred lines of different breeds.
Individual selection Selection on the basis of individual records (as distinct from records of relatives).
Joule Unit of energy. A joule is the work done when the point of application of a force of one Newton is displaced through a distance of one metre in the direction of the force.
Line breeding Inbreeding to a single common ancestor.
Live vaccine A vaccine containing the living pathogen which has been rendered incapable of producing disease but can still induce immunity in the animal or bird.
Locus (plural loci) Position on the chromosome occupied by an allele of a particular gene.
Lux 1 lux = 1 metre candle = 1 lumen/m^2.
Maintenance At the maintenance level of feeding, a bird's requirement for a nutrient or for energy for the normal continuance of vital processes within the body, including the replacement of endogenous losses in faeces and urine and from the skin, is just met so that the net gain or loss of the nutrient from the tissues of the bird as a whole is zero.
Metabolizable energy (ME) The metabolizable energy of the food (ME or M in formulae and equations) is the heat of combustion of the food minus the heat of combustion of the faeces and urine. At the maintenance level of feeding, metabolizable energy, by definition, equals the heat production.
Metabolism The sum of the processes concerned with production and destruction of protoplasm in the course of life. This is commonly measured in terms of rate of heat production in an animal body by the combustion of food, and expressed in units of kilogram joules (KJ) per square metre.
Microclimate The climatic conditions directly surrounding the animal.
Mutation Change in the chemical structure of a gene.
Mutation rate Proportion of genes changed in a given manner each generation.
Nematodes Unsegmented worms some of which parasitise birds and other animals.
Nicking Combination of two lines with unexpectedly good results.
Nitrofurans Synthetic drug used in the control of coli and other bacteria.
Outbreeding Mating of individuals less closely related than the average of the population.
Overdominance Superiority of a heterozygote over both homozygotes.
Performance test A comparison between the individual performance

(production) of a series of prospective parents kept under the same environmental conditions.

Phenotype The appearance and/or performance of an individual.

Phenotypic correlation Correlation between two measurements on the individuals of a population, separable into *genetical* correlation and *environmental* correlation.

Phenotypic variance The variance for individual measurements of a character in a population.

Pleitropy The influence of one gene on more than one character.

Polygenes Genes whose individual small effects (on quantitative characters) cannot be recognised in the phenotype because they are obscured by other variations.

Polypnoea Shallow rapid breathing.

Population A group of interbreeding individuals sharing a common gene pool (e.g. a closed stock, a breed).

Progeny test Evaluation of the breeding value of an animal from the average performance of its progeny.

Prophylaxis The prevention of a disease or control of its spread.

Protozoa Literally 'first animals'. The simplest of all animals which are made up of only one cell.

Proven sire Sires that have shown their ability to pass on productive characters through progeny tests. Sires can be proved for any level of productivity.

Recessive gene A recessive gene is detectable only in the homozygous state.

Regression to the mean The tendency for offspring to produce near to the average for the population.

Relative humidity The amount of water vapour in the air divided by the amount of water the air could contain expressed as a percentage.

Selection Systematic choice of animals in a population as parents for the next generation. (Strictly speaking this is *artificial* selection as distinct from *natural* selection.)

Selection differential Difference between mean of animals selected for breeding and the whole generation to which they belong.

Selection intensity The proportion of the animals produced in a given generation which are selected to be parents of the next.

Sex limited Of characters normally expressed phenotypically in only one sex (e.g. egg production).

Sex linked Of genes located on the sex chromosomes, e.g. silver and gold colour of certain breeds of chicken.

Sib Full brother or sister.

Sib test Evaluation of the breeding value of an animal from the average performance of his or her sibs.

Solar radiation The total electromagnetic radiation emitted by the sun. About one half of the total energy in the solar beam is contained within the visible spectrum from 0.4–0.7 micrometres, and most of the other half lies in the near infra-red, a small additional portion lying in the ultraviolet.

Species A group of animals or plants which have one or more distinct characters in common by which they may be differentiated from other species. Members of the same species are usually capable of producing fertile offspring when mated with other individuals within the species.

Sulphonamides Substances which help to control certain bacterial disease by inhibiting their growth.

Tandem selection Selection for two characters in alternate generations.

Topical application Used in the sense of applying a drug or an ointment to the surface of the body.

Vaccination Inoculation of a bird with attenuated micro-organisms or viruses to enable it to build up immunity to a specific disease.

Viruses Disease-producing agents that are so small that they cannot be seen by normal microscopes. They are the smallest living organisms.

Bibliography

A.T.B., *Candling and Packing Eggs* (Agricultural Training Board, Kent England, 1981)

Collecting Eggs (Agricultural Training Board, Kent, England, 1984)

Depopulating a Poultry House (Agricultural Training Board, Kent, England, 1981)

Handling Turkeys (Agricultural Training Board, Kent, England, 1982)

Killing Chickens and Disposing of Dead Birds (Agricultural Training Board, Kent, England, 1978)

Maintaining Flock Health (Agricultural Training Board, Kent, England, 1983)

Medicating Poultry (Agricultural Training Board, Kent, England, 1983)

Operating and Maintaining Rearing and Laying Cages (Agricultural Training Board, Kent, England, 1982)

Poultry Egg Production – Managing of Breeding Stock (Agricultural Training Board, Kent, England, 1983)

Preparing and Restocking a Poultry House (Agricultural Training Board, Kent, England, 1987)

Taking Blood Samples from Poultry (Agricultural Training Board, Kent, England, 1983)

The Broiler Crop – Crop Timetable (Agricultural Training Board, Kent, England, 1977)

Vaccinating Poultry (Agricultural Training Board, Kent, England, 1983)

ARC, *The Nutrient Requirements of Farm Livestock*, No 1 Poultry (London Agricultural Research Council, 1975)

Balty, J., *Domesticated Ducks and Geese* (Spar Publications, England, 1955)

Farrel, J. and Stapleton, P. (editors), *Duck Production Science and World Practice* (Proceedings of a Workshop at Bogor, Indonesia, Univ of New England, Australia, 1985)

Ferguson, W., *Poultry Housing in the Tropics – Applying the Principals of Thermal Exchange* (Tropical Animal Health and Production 2) pp. 44–58.

Fisher, C. and Boorman, K.N., *Nutrient Requirements of Poultry and Nutrient Research* (Butterworths, London, 1986)

French, K.M., Ritter, L. and Kaufman, Marilyn, *Practical Poultry Raising* (Peace Corps, Washington D.C., USA, 1981)
Gordon, R.F. and Jordan F.T.W., *Poultry Diseases* (Bailliere Tindall, London, 2nd edition, 1982)
Hastings Belshaw, R.H., *Guinea Fowl of the World* (Nimrod Book Services, Hampshire, England, 1985)
Hutt, F.B., *Genetics of the Fowl* (McGraw Hill, New York, London, 1949)
Hutt, F.B. and Rasmussen, B.A., *Animal Genetics* (John Wiley and Sons, New York, 1982)
Johansson, I. and Rendel, J., *Genetics and Animal Breeding* (Oliver and Boyd Ltd., Edinburgh and London, 1968)
Kekeocha, C.C., *Poultry Production Handbook* (Macmillan, London and Basingstoke, 1984)
Kroeshe, D., 'Stimulating Pig Production in Developing Countries (*World Animal Review*, 2, 1973) pp. 15–20.
Kumar, T.V., (Univ of Edinburgh MSc thesis, 1977)
Lyth, P. *Modern Farm Crafts* (Evans Brothers Ltd., London, 1962)
M.A.F.F., *British Poultry Standards* (Butterworths, London, 1971)
Ducks and Geese (H.M.S.O. London, 1984)
Incubation and Hatchery Practices (H.M.S.O., London, 1977)
Intensive Poultry Management for Egg Production (M.A.F.F., London, 1976)
Poultry breeding (H.M.S.O., London, 1960)
Poultry Feeding in Tropical and Subtropical Countries (F.A.O., Rome, 1965)
Poultry Keeping in Tropical Areas (F.A.O., Rome, 1968)
Table Chickens (H.M.S.O., London, 1965)
The Rearing of Pullets (H.M.S.O., London, 1973)
Turkey Production (H.M.S.O., London, 1985)
Mason, I.L. (editor), *Evolution of Domestic Animals* (Longman, London, 1984) pp. 322–324 guinea fowl; pp. 351–356 pigeons.
Matthewman, R., *A Survey of Small Livestock Production at Village Level in the Derived Savanna and Lowland Forest Zones of South West Nigeria* (Monograph, Univ of Reading, 1977)
Mayor, *The Post-mortem Examination of Poultry* (F.A.O., Rome, 1974)
McArdle, A.A., *Poultry Management and Production* (Angus and Robertson, Sydney, 1966)
Nesheim, C.M., Austic, R.E. and Card, L.E., *Poultry Production* (Lea and Febiger 12th edition, 1979)
Nixey, C. and Grey, T.C., *Recent Advances in Turkey Science* (Butterworths, London, 1989)
NRC, *Nutrient Requirements of Poultry* (Washington D.C. National Academy of Sciences, 1981)
Oluyemi, J.A. and Roberts, F.A., *Poultry Production in Warm Wet Climates* (Macmillan, London and Basingstoke, 1979)
Portsmouth, J., *Practical Poultry Keeping* (Spur Publications, Surrey,

England, 1978)

Randall, C.J., *A Colour Atlas of Diseases of the Domestic Fowl and Turkey* (Wolfe Medical Publications Ltd., London, England, 1985)

Richards, S.A., *Temperature Regulation* (Wykeham Publications, London Ltd., 1973)

Romanoff, L. and Romanoff, J., *The Avian Egg* (John Wiley and Sons, New York, 1949)

Sainsbury, D., *Poultry Health and Management* (Granada, London, 1984)

Say, R.R., *Manual of Poultry Production in the Tropics* (CAB International, Wallingford, UK, 1987)

Smit, D.J.V.Z., *Ostrich Farming in the Little Karoo* (Government Printer, Pretoria, South Africa, 1963)

Topps, J.H. and Oliver, J., *Foods of Central Africa* (Research and Specialist Services, Ministry of Agriculture, Harare, Zimbabwe, 1978)

Vietmeyer, N., *Microlivestock* (National Research Council, Washington, USA, to be published)

Warren, A.G., 'Ducks and Geese in the Tropics' (*World Animal Review*, 3, 1972) pp. 35–36.

Wood-Gush, D.G.M., *The Behaviour of the Domestic Fowl* (Heinemann Educational Books Ltd., London, 1971)

Index

aflatoxins, 112, 176
Africa, *3*, *10*, 16
albumen, 42, 149–50
all-in/all-out, 116, 138, 146, 163
alleles, 25, 27, 29, 30, 206
amino acids, *74*
antioxidants, 87–8
appetite; *see* food intake
auto-sexing breeds, *27*, *28*
avian encephalomyelitis vaccine, 133
avian mycoplasma, 146, 171
Aylesbury, *18*, 19, 135

barley, 109
battery cage, *125*
beak-trimming, 118
Beltsville White turkey, 17
biological value, 75, 206
Black Australorps, 16, 183
Blackhead, 143, 166
blastoderm, 150
breast blisters, 139, 142
breeding systems, 23, 37–9
Broad-breasted Bronze, 17
broiler breeders, 5, 7, 44–6
  and diet, *100*
  management, 138–42
  obesity, 133–4
bronchitis, 119, 164, 168
broodiness, 40, 128
bullrush millet, 108
bursitis, 164

calcium, 78–9, 126
candling, of eggs, 158
cannibalism, 118, 141, 188
carcasses, 46, 59, 142, 196, 198, 199
cassava, 110
cattleyards, *144*
cereals, in poultry diet, 107–9, 110
chalazae, 149–50
chickens, species, 12, *13*, *14*
  see also Gallus domesticus
chicks, 116–17, 119
Chinese goose, 20
cholera, 137
chromosomes, 23–4, 25–7
cobalt, 80
coccidiosis, 143, 166, 170–1
coccidiostats, 87, 119, 207
cockerel exchange schemes, 16, 30, 183
controlled environment buildings, 64
conveyor system, 135
copper, 80
copper sulphate, 81
Cornish White, 138
cotton seed meal, 112–13
cross-breeding, 37–9, 46
culling, 35–6, 128–9, 133
curly toe paralysis, 77

day-length, 48, 72, 121
dermatitis, 77
desiccation, of poultry, 50
diet
  for broilers, *100*
  changed by energy specification, *104–5*
  computer formulated, 91–4
  and constraints, 101–2, *103*
  hand calculation, 82–5
  high fibre, 73
  for laying hen, *84*
  mixed, 91
  and Pearson square, 85–6
  for poultry, 106–14
  quality, 102, 106
  and water requirements, 68
  see also cereals; food intake; nutrient requirements; protein requirements
digestive system, 65, *66*
disease control, 7, 140, *164*, *166*

double-yolked eggs, 122
doves, 21–2
dual-purpose breeds, 12, 15, 16
ducks
as egg layers, 135, 137
digestive system, 65, 67
diseases, 137, 164, 172, 173
domestic, 17
feeding, 88
incubation, 161
nutrient requirements, 90
production of, 146–7
scavenging, 190–1
wastage of food, 65, 67, 135–6

earthworms, 185–6, 187
egg production, 39–43
cessation, 48, 69
cleanliness in, 132–3
and food intake, 71
intensive, 115–21
eggs
candling, 158
cleaning, 193
constituents, 148
development, 121–2
fertility, 157
fumigation of, 157
grading, 194
and incubation, 151–6
marketing, 192
nutrient content, 137, 150, *151*
quality, 41, 129
and religious taboos, 1, 194
shape, 41, 129
storing, 132–3, 193
thick white of, 131
weight, 40, 55
Emden goose, 20
encephalomalacia, 77
energy, metabolizable, 69
energy intake, and temperature, 53, *54*, 55
environment, modifying, 60–4
Europe, *3*, *10*

Far East, *3*, *10*
farm populations, *2*
fatty liver kidney syndrome, 77
feather development, 80
feather meal, 114
feather pecking, 117, 118, 121
field ark, *126*
finger millet, 108
fish farming, and poultry, 191
fish meal, 114
flavotoxins, 137
food, containers for, 117, *120*
food conversion, and poultry, 8–9
food intake, 70–1, 72–3
see also diet; nutrient requirements; protein requirements
foodstuffs, nutrient contents of, 92–3, *95*, *96*, *97*, *98*, *99*
fowl cholera, 146, 164, 169
fowl pest, 167–8
fowl pox, 119, 164, 169
fowl typhoid, 171–2
fumigation, of eggs, 157

Gallus domesticus, 5, 7, 15, 67
geese, 17, 20, 67, 189–90
genes, 23–4
genetic correlation, 34–5
genetic improvement, 4–5, 7, 30–1
glucostatic theory, 70–1
gossypol, 113
grading, of eggs, 194
grit, 126
groundnut meal, 112
growth rate, 45–6, 59–60
guinea fowl, 20–1
Gumboro disease, 164

haemoglobin, 80
haemorrhage, 76–7
half-inside/half-outside, *125*
hatching, 158
helminth infection, 173
hepatitis, 137, 164, 172; *see* ducks
heritability, 33–4, 38, 208
heterosis, 38, 39, 43, 44, 45, 208
high fibre diet, 73
Hindus, and egg consumption, 1, 194
hominy chop, 109
Huchzermeyer, 184
humidity, 50, 60
hybrid vigour; *see* heterosis
hygiene, in incubators, 154, 156
hypertension, 80
hyperthermia, 68
hypothermia, 117, *118*

inbreeding, 37
incubation, 152–3, 157, 160, 161
incubators, 7, 153, 154, *155*, 156

inheritance, 23, 24–7, 29–30
iodine, 80
iron, 80
iron oxide, anhydrous, 81

kaffir corn, 118
Khaki Campbell, 17, *19*, 135

Latin America, *3*, *10*
laying birds, 4–5, 43, 119, 121, 128–9
laying houses, *203*
laying phase, 40, 121–2, 126–7
leukosis, 164
lice, 175
Light Sussex, 16, 45
lighting, 7, 72, 126, 141
limestone flour, 81
linseed meal, 113
lipostatic mechanism, 71
liver fluke, 137
Lyallpur Silver Black, 183

maggots, 185, 186–7
maize, 107–8, 109
Mammoth Bronze, 17
manganese, 79–80
manganese sulphate, 81
Marek's disease, 119, 164, 169–70
Matthewman, 184–5
meat meal, 114
medullary bones, 126
Mendelian inheritance, 24, 30
metabolic rate, 50, *51*
milo, 108
mineral pre-mix, *87*
mineral requirements, of poultry, 78–81
mites, 175
mortality, of poultry, 163
Moslems, and killing of poultry, 196, 198
moulting, 69, 128
Muscovy, 17, *18*
mycoplasma, 146, 171
mycotoxicosis, 176

Near East, *3*, *10*
nematodes, 209
Nepal, and semi-intensive poultry, 183–4
New Hampshire, 16, 45, 138, 183
Newcastle disease, 119, 164, 167–8, 178
nicking, 38–9, 209
Nigeria, 16, 179
Norfolk Black, 17
North America, and animal protein, *3*
nutrient content, of various foods, *95–9*
nutrient requirements, 113–14
  for ducks and hens, 90
  for humans, 8, *10*
  for laying fowls and ducks, *91*
  for laying hens, *83*
  of poultry, 4, 67
  for turkeys, *89*
  see also diet; food intake; protein requirements
nutrition, 50–1, 59–60

obesity, in broilers, 133
oil seeds, 111
öoporphyrin, 42, 130
osteoporosis, 76
ostriches, 23
oviposition, *see* laying phase

panting threshold, 52, 57, 68
parasites, 173–6
pasteurellosis, *164*, 169
Pearson square, 85
pebble yards, *125*
Pekin, *18*, 19, 135
perosis, 79
pheasants, 22
phosphorus, 78, 79
pigeons, 21–2
pleiotropy, 34
plucking, 198
pole barns, *144*
population genetics, 30–1
poultry, indigenous, 46
poultry houses, 61, 62–4, 162–3
poultry production 7–8, 11
poultry products, and marketing, 192, 194–6
preselected strain crossing, 38–9
profits, from poultry, 200–1
prophylaxis, 165, 210
protein, from animals, 1, *3*, 113–14
protein requirements, of poultry, *9*, 74–5
  see also diet; food intake; nutrient requirements
pullorum disease, 166, 171–2

quail, 22

range rearing, of turkeys, 143
records, for poultry keeping, 202, *203*, *204*, *205*
respiratory disease, chronic, 171
Rhode Island Red, 16, 183
rice, 109, 191
rickets, 76
Roman goose, 20
roundworms, 173

salivary glands, 76
salmonella, 114, 177
salt, 80
sanitiser, for eggs, 193
scavenger poultry units, 179–83
scavenging flocks, 1
  and diet, 184–5
  and disease, 176–8
  management, 187–9
  in Sudan, 179–80
  in West Africa, 3–4, 179
selection, 31–3, 36–7, 39
semi-intensive production, 183–4
sex, inheritance, 24–6
sex cells, 23–4
sexing, 159
shell
  colour, 42, 130
  formation, 76, 78–9
  membranes, 148–9
  quality, 41–3, 130
  weight, 56
skip-a-day feeding, 73
slaughter, of birds, 196, 198
social organisation, of poultry, 115–16
sorghum, 94, 108
soya bean, 111, 112
space requirement
  for chickens, 119, 123
  for turkeys, *144*, *145*
spots, blood and meat, 43
stress, 50, 115–16
Sudan, 179–80
sugar, 111
sunflower seed, 113

taints, 132
tandem method, 35–6, 211
tapeworms, 173
tapioca, 110
temperature, 50–1, 52–60
  for broilers, 141
  and carcass composition, 59
  for chicks, 116
  effective, 47–8
  and egg weight, 55
  and energy intake, 53, 55, 59–60
  and shell weight, 56
  and stress, 50
  and water intake, 56–7, 68
thermal neutrality, zone of, 50–1
thermostatic theory, 70
ticks, 175
Toulouse goose, 20
traits, selection for, 32–3, 34–6
turkeys, 17
  breeding, 134
  in Britain, 195
  domesticated, 16
  feeding, 88
  incubation, 161
  nutrient requirements, *89*
  producing, 142–3, 146

vaccination, 119, 133, 162, 165, 211
vasodilation, 52
vent-sexing, 7, 160
ventilation, in poultry house, 63
vertical integration, 8
viability of hens, 40–1
vitamins, 75–8
vitelline membrane, 150, 212–3

Wallace, 38
wastage, with ducks, 65, 67, 135–6
water requirements, 56–7, 67–9, 126, 187
water glass, 132–3
welfare, of egg laying poultry, 115–16
West Africa, and scavenger flocks, 3–4, 179
wheat, 108–9
White Cornish, 45
White Leghorns, 183
White Plymouth Rock, 138
White Rock, 45

yolk, 43, 130–1, 150
yuca, 110

zinc, 80, 81
zinc carbonate, 81